ENVIRONMENTAL MANAGEMENT

Environmentally-sound Water Management

Edited by
N. C. THANH
ASIT K. BISWAS

DELHI
OXFORD UNIVERSITY PRESS
BOMBAY CALCUTTA MADRAS
1990

Oxford University Press, Walton Street, Oxford OX2 6DP
New York Toronto
Delhi Bombay Calcutta Madras Karachi
Petaling Jaya Singapore Hong Kong Tokyo
Nairobi Dar es Salaam
Melbourne Auckland
and associates in
Berlin Ibadan

SBN 0 19 562744 X

Photoset by Indraprastha Press (CBT), New Delhi 110002
Printed by Rekha Printers Pvt. Ltd, New Delhi 110020
Published by S. K. Mookerjee, Oxford University Press
YMCA Library Building, Jai Singh Road, New Delhi 110001

Contents

List of Tables and Figures

Foreword

In the last two decades, much has been written about environmentally-sound development. More recently the term sustainable development was coined. While everyone agrees on the need for development that sustains and is sustainable, experience shows that translating the term from a concept into everyday, operational priorities remains a very difficult task.

Clearly, development must be environmentally sound to ensure that development process does not destroy the resource base upon which future growth depends. This is a fundamental tenet of development. Some 16 years ago, during the World Food Conference in Rome, I called this approach 'development without destruction'. More recently, such seminal reports as UNEP's *Environmental Perspectives to the Year 2000 and Beyond,* the *World Conservation Strategy* and *Our Common Future* have sharpened our understanding of environmentally sound and sustainable development.

Environmental costs of development will only be minimized when development decisions are guided by ample knowledge and information. When development decisions were made—say some 40 years ago—decision-makers were not fully aware of the long-term environmental and social impacts of their decisions. Not surprisingly, decades later we are still paying the price of some earlier decisions, in terms of destruction of the natural resource base and environmental degradation.

Now, with our knowledge base substantially enhanced, we must learn from our past mistakes, and avoid making the same errors again. Environmental considerations need to be integrated rationally and efficiently in natural resources planning and development, thereby minimizing and halting the destruction of our planet's resource base.

We have made reasonable success in intergrating environmental considerations at the project level and some success at the sector level. Unfortunately, we have not succeeded in designing means of integrating environmental concerns effectively at the national plan and policy levels, except in very general terms. Currently, in spite of the rhetoric, appropriate operational methodology for such integration simply does not exist. Unquestionably, increased attention must be paid to improving our knowledge on how to integrate properly environmental concerns at the national plan and policy levels. Water resources development is no exception to this general state of affairs.

Water is the lifeblood of all development. Not surprisingly, all the major ancient civilizations developed on the banks of major rivers like the Nile, Indus, Tigris and Euphrates. Water was considered to be an important element in most early religious and classical texts.

The role of water in the development of developing countries is no less important today than in ages past. The food and energy crises that most developing countries face today will not be solved on a long-term, sustainable basis without efficient water management policies. Environmentally-sound water management is an absolute necessity if the ravages of floods and droughts are to be significantly reduced. The prolonged tragic droughts in Africa and floods in many parts of the world have sadly confirmed this fact. And the threat of climate change and global warming makes sustainable water management policies imperative, particularly given long lead times in resource planning.

Water development projects, like any other type of resource development, have both positive and negative impacts. The primary task facing planners and managers is to maximize the positive impacts of such development and minimize the negative ones. Recent discussions of these impacts have concentrated primarily on the negative impacts. What needs to be better understood are positive impacts of sustainable water management. Simply put, unless positive aspects *significantly* exceed negative impacts, development projects should not be built.

Let us consider the case of the Aswan High Dam in Egypt, which is included in this book as a case study. Among the many adverse environmental impacts attributed to this dam are salinity growth, increases in schistosomiasis, erosion of the Nile Delta and loss of some of the Mediterranean fishing grounds. But what is worth noting is that even after the construction of the Aswan High Dam, availability of per capita arable land now in Egypt is still around 0.07 ha, the lowest of any country in Africa. There is no argument: the dam's contributions to substantially increasing food production and hydropower generation have made an enormous difference to the quality of life of the Egyptian people. Thus, the real question is not whether the Aswan High Dam should have been constructed, but rather what steps should have been taken to anticipate and address the adverse environmental impacts, and to ensure they were reduced to an acceptable minimum. And in this particular case the adverse impacts were all anticipated and plans set for addressing them but international political situations delayed the implementation. So the overall cost of the negative impacts and of their redressing skyrocketed like any other cost in the world today.

What is thus necessary is a *balanced* environment–development philosophy, within which environmentally-sound water development could occur.

I am pleased that Dr N. C. Thanh (Director of Asia Division, CEFIGRE) and Professor Asit K. Biswas (President of the International Water Resources Association) have now prepared a study examining and elaborating such a balanced environment–development perspective. It is the first book that I know which evaluates environmentally-sound water management, and deals in an integrated manner with the issues related to water development in developing countries. This book fills a long-standing need in this area, and is a welcome addition to the growing body of literature on environmentally-sound and sustainable development.

MOSTAFA KAMAL TOLBA
Executive Director
United Nations Environment
Programme, Nairobi,
Kenya

Preface

Throughout human history, water has always been considered to be an important requirement for human welfare and economic development. The role of water became even more important during the past two decades for a variety of reasons. First, severe droughts in many parts of the world, and especially in Africa, contributed to a major food crisis. In Africa, per capita food production declined steadily during this period as population increased far more rapidly than agricultural production. Increases in agricultural production are only possible through further expansion of cropped areas and increases in agricultural yields. Both of these are only possible provided reliable sources of water are available.

Secondly, prices of fossil fuels, especially oil, increased steadily from 1973. This focused the attention of many governments to another major source of energy, i.e. hydroelectric power, which is not only a renewable source of energy but is generated without any consumption of water. Thus, the possibility of developing major sources of water that could be used to expand irrigated areas and simultaneously generate hydroelectric power, became an attractive proposition to many countries.

Thirdly, the United Nations convened a world Water Conference in Mar del Plata, Argentina, in March 1977, at a high decision-making level. This conference, the first of its kind ever held in the area of water, sensitized the world community on the importance of water for development, especially for developing countries. The conference unanimously agreed on an Action Plan. An important component of the Mar del Plata Action Plan was to declare the 1980s as the International Water Supply and Sanitation Decade. This declaration heightened the importance of availability of clean water to every citizen of the world.

Another issue worth noting is that as the interest in water development in developing countries accelerated during the last two decades, so did the concern with the environmental impacts of development projects. As the decade of the 70s progressed, the interrelationship between environment and development came into sharper focus and its importance became increasingly clear to policy-makers and the general public. It was gradually accepted that environment and development are two sides of the same coin. Much of the credit for clarification of this concept as well as its acceptance by developing countries must go personally to Dr Mostafa Kamal Tolba, Executive Director of the United Nations Environment Programme, who made it one of his priority areas. As a result of such efforts, environmental concerns are now being incorporated to the extent possible into development projects of most developing countries, and water resources development is no exception.

All development projects have environmental, economic and social consequences, some beneficial and others adverse. In the nineties, the main challenge facing the water resources profession is how to maximize all the positive impacts of any water development present, planned or already operational, and minimize the adverse impacts. It is now evident that water resources projects can be better planned and managed to ensure more reliable water availability and efficient water use in the agricultural sector, mitigate flood damage, and control water pollution, and simultaneously reduce adverse environmental and social impacts such as prevention of development of waterlogging and salinity, reduction in the spread of water-borne diseases and proper resettlement of displaced people. Environmentally-sound water management should be able to address and resolve all these issues simultaneously.

While no sane person will currently question the importance and desirability of environmentally-sound water management, it is really a sad commentary on our present status that a book on this important subject simply does not exist. In order to fill this critical gap, International Training Centre for Water Resources Management (CEFIGRE) in Sophia Antipolis, France, undertook to develop this text in close collaboration with and support of the United Nations Environment Programme (UNEP) and International Water Resources Association (IWRA). The present book is a direct result of this effort. While the main objective of this book is to promote environmentally-sound water management in developing countries, we are confident that the water professionals in other countries will find it useful as well. CEFIGRE, UNEP and IWRA plan to use this book as the basic text for a series of training courses on environmentally-sound water management for developing countries.

The editors wish to express their most sincere appreciation to

Dr Mostafa Kamal Tólba, Executive Director of the UNEP, for his constant encouragement to develop this book. In spite of his numerous international commitments, Dr Tolba kindly agreed to write a Foreword to the book, for which we are truly grateful. We also wish to acknowledge the support we have received from Dr J. Balek, Dr H. El-Habr and the late Dr L. David (all from UNEP) and our colleagues from CEFIGRE. Without such support, this book would not have been possible.

N. C. THANH
Director, Asia Division,
International Training Centre for
Water Resources Management
(CEFIGRE), Bangkok, Thailand

ASIT K. BISWAS
President, International Water
Resources Association,
Oxford, England

List of Contributors

N. C. THANH
CEFIGRE, France/Thailand

D. M. TAM
Cowater International, Canada

ASIT K. BISWAS
International Water Resourses Association,
Oxford, England

JANUSZ KINDLER
Institute of Environmental Engineering,
Warsaw Technical University, Warsaw, Poland

WANCHAI GHOOPRASERT
Provincial Waterworks Authority, Thailand

M. B. PESCOD
Department of Civil Engineering,
University of Newcastle upon Tyne,
United Kingdom

R. MACKAY
Department of Civil Engineering,
University of Newcastle upon Tyne,
United Kingdom

MAHMOUD ABU ZEID
Water Research Centre, and
International Water Resources Association,
Cairo, Egypt

CHAPTER 1

Water Systems and the Environment

N. C. THANH and D. M. TAM

FUNDAMENTALS OF ECOLOGY: HUMANS IN THEIR ECOSYSTEM

The word 'ecology' is derived from the Greek *oikos*, meaning 'house' or 'place to live'. Ecology is the study of interrelationships between living organisms and their environment. It is concerned with the biology of groups of organisms and with functional processes in the entire environment which include the air, soil, freshwater bodies and the seas.

Ecosystem is a unit comprising organisms and processes. Its state can vary. It can be as small as a pond, or as large as a lake, a forest, or even larger such as a catchment area of a large river. Generally an ecosystem consists of (i) *inorganic substances* such as water and carbon dioxide; (ii) *organic compounds* such as proteins, carbohydrates and fats; (iii) *climatic regime* such as temperature, wind and precipitation; (iv) *producers* (mainly green plants) which are able to build their bodies using simple inorganic elements; (v) *consumers* (humans and animals) which are not able to use simple elements but must consume organic compounds of other organisms to build their bodies; and (vi) *microconsumers* or *decomposers*, chiefly bacteria and fungi, which break down complex compounds into simple elements that plants can absorb. The first three groups constitute the *non-living (abiotic) components*, whereas the last three constitute the *living organisms (biomass)*. Each of the three living organism groups have distinct general functions in the ecosystem: *production*, *consumption* and *decomposition*, respectively. All three must exist together in a stable ecosystem, and all three must live in harmony with the three abiotic component groups. For instance, without the decomposers organic matter will accumulate in dead bodies which plants cannot use, and so no new life can start.

In order to understand the importance of environmental conservation and mitigation of potential adverse impacts on the environment, the following main characteristics of a stable, healthy ecosystem should be noted:

- No organism can live by itself but rather each depends on others for survival. In turn, each of the living organisms plays a useful role in the ecosystem; all species constitute the wholeness of the biosphere. For this reason, one organism will affect the future of others.
- There are mechanisms within individual organisms and in the ecosystem for self-maintenance and self-regulation, and this makes an ecosystem stable. Examples in the ecosystem are numerous: the balance among grasses, rabbits and wolves, or among plankton, herbivorous fish and carnivorous fish. This mechanism (*homeostasis*) develops only after long periods of adjustments. It may have limitations beyond which the system may collapse. After that, it may take some time to recover, if it can ever recover, or the process could be irreversible.
- For the various elements necessary for each organism to grow and reproduce, the single element closest to the critical minimum will be a factor limiting the growth and reproduction of that organism. If everything else is plentiful in water, but light is limited (e.g. by turbidity) then light could be the limiting factor for algae; all other elements may not be useful. This is the *Liebig's Law of the Minimum*.
- Each organism has its lower and upper limits of tolerance for living conditions, such as for temperature. Also, 'too much of good things' will become harmful. This is the *Shelford's Law of Tolerance*.

Stated simply, as such these principles do not reveal all the intricate features in the ecosystem. However, these (and others) are important governing rules which should be carefully considered and applied to human activities that may affect the environment.

The Hydrological Cycle

The hydrological cycle is basically the circulation of water within and around the earth as water is converted from one form to another and transported from one place to another. The driving forces of the hydrological cycle are the sun's energy and the earth's gravity. Most active phenomena occur in the oceans, where each year about 453,000 km^3 of water evaporates to the atmosphere. Water vapour in the atmosphere condenses and precipitates on the surface in various forms such as rain, snow, sleet and hail. About 90 per cent of the 453,000 km^3 of water returns each year to the oceans; the remaining 41,000 km^3/year precipitates on land together with 72,000 km^3/year of water evaporated from land masses and falls back. Water then follows three different paths, namely infiltration or percolation, evaporation or evapo-transpiration, and runoff.

Infiltration/Percolation. A major portion of precipitates infiltrates below the surface and into pores and fractures of geological formations called aquifers. Underground water eventually discharges to surface waters through springs or groundwater outflows in rivers or lakes.

Evapo-transpiration. Through evaporation from the earth's surface and evapo-transpiration from plants, water constantly returns to the atmosphere as vapour, which then circulates around the earth and precipitates again. While evaporation is an important process in the hydrological cycle, plants magnify the effect greatly by transpiring considerable masses of water. In addition, their root systems accelerate the upward movement of water against the force of gravity. In this way, transpiration by vegetation vaporizes large quantities of water: a hectare of forest in temperate climates can evaporate 20–50 tonnes of water per day (Ramade 1981). The combined effect of evapo-transpiration is thus an important factor in the hydrological cycle of land-based ecosystems. In tropical forests, such as those in the Congo and Amazon basins, the majority of rainwater originates from evapo-transpiration; only 20 per cent comes from the seas.

Runoff. About 20 per cent of precipitation reaches springs, lakes and rivers. A major part of surface flows is in sparsely populated areas such as the Amazon River Basin, leaving only 9000 km^3/year for human use. Eventually 41,000 km^3/year of surface water on land returns to the oceans to complete the hydrological cycle. While the movement through some parts of the hydrological cycle may be relatively rapid (for instance, water vapour on the average remains in the atmosphere for 10 days), water moves very slowly underground. Complete recycling of groundwater may be measured in thousands of years. This fact is seldom understood by groundwater users who 'mine' these 'fossil aquifer' at a rate much higher than the replenishment rate (see section 'Excessive water withdrawal and its impact' below).

About 97.3 per cent of the earth's water is in the oceans and bodies of saline water. Of the remaining 2.7 per cent, which is freshwater, 2.1 per cent is tied up in the polar ice caps and in glaciers, leaving only 0.6 per cent to circulate. Groundwater and soil moisture constitutes 22.4 per cent of the global freshwater, but two-thirds of the groundwater reserve lies below 800 metres depth and is beyond human capacity to exploit. For all of their livelihood purposes, humans must depend upon the remaining approximately 0.2 per cent found in freshwater lakes, rivers and up to 800 metres in the ground. Contrary to popular belief, the amounts of freshwater in lakes and rivers are very small, constituting about 0.009 per cent and 0.0001 per cent, respectively, of the world's total water. Moreover, the distribution of these tiny portions are grossly uneven. For example, nearly 50 per cent of Africa's total surface water resources are in the Congo River basin, and Lake Baikal alone contains 10 per cent of the total freshwater stock of the whole planet! Coupled with all abuse created together with the population growth and the increase in per capita water demand, water crisis is a reality.

Even though the major rivers of the world have very large discharges, the amount of water contained in these rivers at any given time is small. For instance, if all the water of all rivers of the world could be pooled together at any given time, the resulting lake would still be smaller than Lake Ontario. Thus, in water resources development planning and design, the dynamic flow over time is a more important factor than the static holding volume.

From the above, we can see that understanding our ecosystem and the hydrological cycle is the essential beginning of water resources development planning. Humans should not disrupt the orderly functions of the hydrological cycle, or of hydrogeochemical cycles in general. It should be noted that the hydrological cycle is only an integral part of hydrogeochemical cycles, in which water, air and soil are interrelated. Human activities have caused profound effects on the hydrogeochemical cycles. The most sensitive components to these effects are the atmosphere, the living biomass (mostly forests) and ground and surface freshwaters (Stumm 1986).

WATER USES

Figure 1 indicates that global water use during the period 1950–80 increased by more than three-fold, mainly because of the fast increase in per capita water use. During the same period, world's population increased by 80 per cent only, which shows the degree of stress humans have put on the environment.

Agricultural Use

Agriculture is the largest user of water, accounting for nearly three-quarters of the total global consumption. Large portions of water used are lost through evapo-transpiration and seepage. Accordingly, more water must be provided onto the fields than what is required by the crops. Often more than half the water withdrawn for irrigation does not reach the field. Approximately one-third of the agricultural production comes from the 18 per cent of the world's cropland that is irrigated. It is estimated that by 1990, the total area irrigated in the world will be some 273 million hectares, of which 119 million

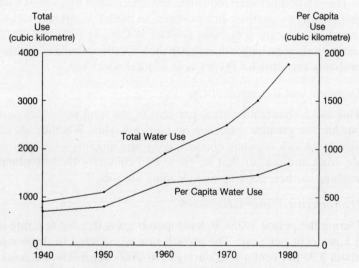

FIGURE 1. World water use, total and per capita, 1940–80

(Source : USSR Committee for International Hydrological Decade and Worldwatch Institute)

hectares will be in developing countries. By the turn of the century, the volume of irrigation will be increased by an estimated rate of 25–30 per cent over the 1980's level.

Industrial Use

Industry is the second major user of water, accounting for about a quarter of total water used worldwide. The largest industrial water users are power plants (both nuclear and fossil-fuel) since large quantities of water is used for producing steam to drive turbo-generators and for cooling. However, only a small portion (2 per cent in the USA) of water withdrawn is consumed. The volume of water withdrawn by these plants are of less concern than the discharge of heated water back to the environment.

Of the remaining water withdrawals for industry, two-thirds go to just five industries: primary metals, chemical products, petroleum refining, pulp and paper manufacturing, and food processing. Developing countries use about 20–40 m^3/person-year as compared with 2300 m^3/person-year in the USA.

The relation between industrial and agricultural uses varies from one country to another. For example, in the USA agricultural and industrial uses are somewhat similar, in Canada industrial water use is significantly higher than that in the USA, but in India agriculture probably accounts for 90 per cent of total water use.

Domestic and Recreational Use

This use accounts for only 6 per cent of the total water demand, but has the greatest influence on human health. Whether we are healthy or sick depends on the quality and quantity of water that we drink and use, or that we come into contact with for bathing, washing clothes, cooking and cleaning utensils.

Hydroelectric Power Generation

During the period 1925–70, hydropower grew steadily at a rate of 6.5 per cent per year. The growth rate is expected to decline to about 3.75 per cent a year during 1970–2000. Hydroelectric power currently accounts for 70 to nearly 100 per cent of all electricity generated in Brazil, Canada (where electricity is called simply *hydro*), Morocco, Norway and Sri Lanka. In general, hydropower potential in many developing countries is still much under-developed: while these countries possess 50 per cent of the world's hydropower resources, they have exploited only 10 per cent of their potential.

WATER RESOURCES DEVELOPMENT AND THE ENVIRONMENT

Water resources development activities are not new in human history. As early as 3000 BC the Egyptians had developed intricate water networks, especially for irrigation systems. In China, the Huang He (Yellow River) has been used for irrigation for nearly 2500 years, and the origins of the Grand Canal date back to the same period. Water resources development activities along the Tigris, Euphrates and the Indus Rivers also date back thousands of years. However, only during the last 4–5 decades both the magnitude and number of water projects have increased significantly, with concomitant impact on the environment. The various aspects of environmental impacts will be discussed in the following sections.

Environmental Inventory

The environmental inventory consists of physical components present in the ecosystem of a project area that need to be assessed both qualitatively and quantitatively. Table 1 (adapted from Roser 1976) serves as a checklist of those components in an environmental inventory that should be considered in planning water resources development projects. The actual list would depend on specific site conditions.

Environmental Parameters

Environmental parameters (also called 'attributes') are variables that can be quantified or clearly described and that represent or measure the characteristics of the environment. In short, they serve as 'yardsticks' of the environmental inventory listed above. For example, as shown in Table 2 (adapted from Jain *et al.* 1981; Rau and Wooten 1980), 21 parameters can be used to describe the characteristics of the water resources components listed in Table 1.

AREAS OF ENVIRONMENTAL IMPACTS

Although any environment impact could be either beneficial or adverse, in environmental analysis, impacts are historically considered only to be of adverse type caused by our developmental activities. Impacts can be generally categorized as primary, secondary or tertiary. *Primary impacts* are those caused directly by project inputs such as loss of forests, changing of a river regime due to the construc-

TABLE 1. Environmental inventory

Topographic and scenic components	Vegetation resources
Basic geology	Virgin timber stands
Prominent elevations	Rare timber remnants
Caves and sinkholes	Reforestation activities
Cliffs and outcroppings	National forests
Stone and mineral collection areas	Orchards
Ore outcroppings	Nurseries
Agricultural activities	Game inhabited areas
Soil conservation activities	Important habitats
	Endangered plant species
Climatic regime	*Historical and cultural resources*
Rainfall	Historical structures
Temperature	Museums
Wind	Temples and shrines
Evaporation	Cemeteries
Humidity	Traditional cottage industries
Solar irradiation	*Wildlife resources*
Potential for natural disasters,	Biological diversity
e.g. flashfloods	Animals
	Fish
Water resources	Birds
Drainage patterns	Breeding grounds
Significant streams	Sanctuaries
Groundwater	Endangered species
Lakes and ponds	*Recreational facilities*
Beaches	Hunting areas
Wetlands	Water-related activities
Dams and reservoirs	Camp sites, picnic grounds
Canals and locks	Resorts, tourist accommodations

tion of a dam. As such, primary impacts can be attributed directly to a project activity. They are usually easy to measure. *Secondary impacts* are those caused by project outputs such as water flow regulation and channelization. In other words, they are indirectly attributed to the project activity. If one of the project outputs is availability of irrigation water, secondary impacts could be the effects on the fertility of the irrigated soil. Secondary impacts could be more severe than primary impacts and, unfortunately, often more difficult to predict and measure. Secondary impacts in turn may lead to tertiary impacts.

It should be noted that the distinction between primary, secondary

TABLE 2. Environmental parameters

Water
1. Aquifer safe yield
2. Flow variations
3. Chlorides
4. Oil and grease
5. Suspended solids
6. Acidity/alkalinity/pH
7. BOD and/or COD
8. Dissolved oxygen
9. Nutrients
10. Trace elements, e.g. Mn, Mg, Bo, Se
11. Pesticides
12. Heavy metals
13. Other toxic substances, e.g. PCBs
14. Faecal coliforms
15. Faecal streptococci
16. Temperature
17. Radioactivity
18. Aesthetics: colour, odour, taste

Air
19. Particulates
20. Sulphur oxides
21. Hydrocarbons
22. Nitrogen oxide
23. Carbon monoxide
24. Photochemical oxidants
25. Hazardous chemicals, e.g. lead
26. Odours
27. Visibility

Noise
28. Noise levels
29. Physiological effects
30. Psychological effects
31. Communication effects
32. Performance effects
33. Social behaviour effects

Flora/Fauna
34. Species checklist/counts
35. Species diversity
36. Growth rate
37. Migration patterns

Land
38. Soil types
39. Soil composition: nutrients and trace elements
40. Soil physical characteristics: bulk density, permeability, etc.
41. Soil stability
42. Natural hazards; seismicity, floods
43. Salinity intrusion

Socio-cultural environment
44. Infrastructures and basic services: water supply, sanitation, health care, school, etc.
45. Housing conditions
46. Socio-cultural and ethnic characteristics
47. Community life; community organization, self-help activities, places to meet, festivals
48. Religions
49. Demographic characteristics
50. Women's role in household and in development

Economic environment
51. Labour and wages structures
52. Industrial outputs
53. Total and unit yields of agricultural and fishery products
54. Income: level, source and distribution
55. Land tenure and value
56. Landownership

People
57. Health indices, e.g. life expectancy
58. Nutritional status
59. Eating habits
60. Hygiene practices
61. Health status: diarrhoeas, parasitic diseases, other infectious diseases, e.g. tuberculosis
62. Settlement and migration patterns

and tertiary impacts could often be arbitrary. In the ecosystem, impacts are usually complex and one impact may lead to another, resulting in chain-actions (e.g. deforestation could contribute to increased reservoir siltation which could lead to a loss in downstream fishery, causing malnutrition which in turn may increase sickness) and a major impact may often be due to a combination of factors. The causal linkages between impacts may not be direct or clear-cut. Various types of water-related activities can cause beneficial or adverse impacts on the environment. These activities may include land clearance, construction, water impoundment, water channelization, flood land alteration and changes in land-use patterns.

Impacts could also be conceptually divided into two broad categories: short-term and long-term. Short-term impacts occur during the planning, construction and immediate post-construction phases. Longer-term impacts stem from the presence of large man-made lakes, development of perennial irrigation instead of seasonal irrigation, alteration in the ecosystem of the area, and the changing socio-economic situation.

SHORT-TERM ENVIRONMENTAL IMPACTS

Many adverse impacts—such as noise, soil disturbance, air pollution, disruption of transport—are inevitable results of activities during the construction phase. These are usually tolerable. However, major short-term social and environmental impacts are often caused by the large workforce at the construction site. Water resources development projects requiring major construction create new employment opportunities, and so new workers of different categories move into the areas near construction sites in large numbers. If these projects are undertaken in remote, undeveloped regions, they lack suitable housing, sanitation and other facilities. Construction of such supporting infrastructure invariably have environmental implications. Even when they are not in remote regions, host communities are seldom able to absorb large numbers of immigrant workers without encountering serious environmental and social problems.

In developing countries labour-intensive technology is often used to construct large structures, which means that a large number of unskilled and uneducated labourers arrive at construction sites in search of employment. The construction of large water development projects often lasts over a decade, and the daily labour force engaged fluctuates greatly. Under these difficult circumstances and in view of inadequate medical assistance in most developing countries, it is impossible to provide appropriate medical and sanitation facilities to most of the workers.

The presence of large numbers of workers in unhygienic, dilapidated quarters is conducive to the prevalence of disease: tuberculosis, viral hepatitis, meningitis, etc. Furthermore, in some countries like Pakistan, groups of labourers exist who travel all over the country working from one construction site to another. Such movements tend to increase disease transmission rate.

LONG-TERM ENVIRONMENTAL IMPACTS

Physical Impacts

Impacts on Water Quality. Water quality is a very important consideration for all water development projects as it affects all aspects of water use—for humans, for animals, for crops, and even for industry. With regard to water quality in reservoirs and lakes, an understanding of stratification and its effects are essential.

In temperate climates, reservoirs as well as natural lakes are subject to a seasonal cycle of changes in terms of thermal stratification. In the *fall overturn* the warmer top layer (*epilimnion*) become cooler or at least as cool as the bottom layer (*hypolimnion*). As a result, the water of the reservoir mixes thoroughly, and dissolved oxygen and nutrients in the epilimnion and hypolimnion are interchanged. In the *spring overturn*, as the temperature of the epilimnion rises from 0 to 4 °C at which water becomes heaviest, it sinks to the bottom, taking with it dissolved oxygen. Displaced bottom waters rise to the surface bringing with it nutrients deposited on the bottom back to the surface. These biannual mixing processes reduce the extent of thermal stratification and contribute to the maintenance of higher water quality.

By contrast, since the surface temperature of reservoirs in tropical and sub-tropical climates seldom (if ever) falls below 4 °C, thermal stratification becomes a stable year-round phenomenon. Consequently, the hypolimnion is consantly depleted of its dissolved oxygen while nutrients are not recycled. This problem is especially serious for deep reservoirs. Thus, while the same physical structure yields benefits under temperature conditions, it may create serious problems in hot climates. This is an example showing that a knowledge of specific tropical conditions as different from what has been well-understood in temperate climates is vital. Attempts to improve reservoir water quality by releasing hypolimnion water which has a low temperature and little or no dissolved oxygen could adversely affect the ecosystem of the downstream areas. Impoundment of water over long periods enhances settling of suspended solids, hence water quality may be improved in terms of turbidity. However, the deposition of mineral and organic matter may cause problems by silt accumulation and the depletion of dissolved oxygen in the water. The problem of siltation has sometimes been underestimated in large projects, requiring costly remedial measures (Sundborg and Rapp 1986; Wu 1986).

Water below a spillway may become supersaturated with nitrogen as a result of turbulences. Nitrogen contents of 130 per cent have been recorded (normal saturation is considered to be 100 per cent); while the mortality threshold for some fish species appears to be about 120 per cent (ICOLD 1980).

In arid zones the waters often have high salt content. With high evaporation and low precipitation regime, the salinity in impounded water may eventually become exceedingly high.

The nutrient level in water often increase significantly owing to decaying matter in the flooded land and leaching from the ashes of burned vegetation. This may engender eutrophication as indicated by abundant algal growth, higher turbidity, lower dissolved oxygen, and aesthetic problems. Flooding of tropical forest (instead of cutting trees which are removed as timber or fuelwood) may also produce hydrogen sulphide in impounded water, which imparts aesthetic problems for human consumption, causes toxicity to fish, and corrodes metal parts (such as turbines) that come into contact with the water. Studies in Surinam have shown that the cost of turbine repairs corroded by hydrogen sulphide as a result of forest flooding exceeds the expense of deforestation (ICOLD 1985).

Impacts on Groundwater. Irrigation projects may have impacts on groundwater in various ways (Verma 1987). Especially in arid regions, more irrigation water than is needed for evapo-transpiration must be applied to soil to avoid accumulation of salts in the root zone. As a result, groundwater is easily contaminated by fertilizers and pesticides percolating from irrigation water. Considering that there are about 225,000,000 irrigated hectares in the world, groundwater contamination by irrigation projects could be indeed extensive. Irrigation schemes in Punjab, Pakistan, have in their first 10 years of operation, raised the water table 7 to 9 metres above the long-term level.

Large water storage may raise the groundwater level which contacts permeable and polluted beds, thus affecting the quality of groundwater. Once the groundwater table is within several centimetres from the surface, water begins to move upwards by capillary action, pulled by the dry air and soil above. As water reaches the surface and evaporates, a deposit of salts is left behind. Agricultural, industrial and human settlement, development around the new water project area may exacerbate the problem of water quality deterioration.

Impacts on Soil. On the global scale, 46 per cent of the earth's surface affected by soil degradation hazards can be directly related to water. These hazards are roughly estimated as: erosion, 22 per cent; waterlogging and flood damage, 8 per cent; salinity and alkalinity, 5 per cent; and frost, 11 per cent (Biswas 1978). Building a dam over a river deprives the downstream land of nutrients normally brought by the river in the form of silt. Since the Aswan High Dam was built over the Nile, the loss of the nitrogenous components of the silt trapped in the reservoir has been equivalent to 13,000 tonnes of lime nitrate per year. As the Nile no longer carries much silt, its delta some 1000 km away has been eroded, because the soil carried out is no longer replaced by silt brought from upstream. There is still another secondary impact on soil loss. Before the Aswan Dam there was a thriving small-scale industry making bricks from the silt dredged from the canals. In the absence of such silt, these industries started to use the top soil near the canals to make bricks, thus contributing further to the problem of loss of productive soil (Biswas 1978). Egyptian researchers have succeeded in making bricks out of sand, but it took considerable time to persuade the construction industry to switch from silt to sand bricks. On the positive side, however, a decrease in silt loads has reduced the cost of dredging canals.

Soon after the construction of a dam, water is stored to build up water level in the reservoir. This deprives the downstream area of a constant supply of freshwater without which saltwater could intrude into low-land or estuarine areas. Later on as the reservoir fills up, the annual release of water from the reservoir may be lower than that in the pre-construction period, because of higher water loss through evaporation from the reservoir. Thus, freshwater flow to downstream regions is not only substantially cut off during construction but gets reduced *forever*. Saltwater intrusion into low-land areas could become worse with time. This problem has been encountered in the Srinakarin Dam of Thailand where unexpected successive periods of droughts after construction made the reservoir build-up time much longer than expected. Salinity from the Gulf of Siam wiped out some of the richest farmland in the country and destroyed coconut plantations and lychee orchards in coastline Samut Songkhram Province (Tuntawiroon 1985). The affected plantation farmers had to leave their native province to look for other jobs.

Degradation of soil fertility by waterlogging and salinity development is a common, but one of the most serious, impact of irrigation. Increase in irrigation without proper drainage facilities usually leads to increase in soil salinization and alkalization as water evaporates, leaving behind salt particles on top soil layers, causing damage to the vegetation growing on this soil and imparting imbalance in the ecosystem. This problem is more acute if groundwater is used extensively for irrigation, especially when the groundwater has a high dissolved solids content. This occurred in Pakistan during the 1960s when 24,280 hectares of fertile cropland in the country was lost every year (Biswas 1980). Salinization is undoubtedly one of the major environmental impacts of irrigation. Soil salinization and soil erosion due to irrigation practices are becoming serious problems throughout Canada's agricultural areas, particularly in the dry prairie environment (Pearse *et al.* 1985). It is estimated that between 1985 and 2000 the irrigated land area of the world will double, and the problems of salinization and alkalization will increase proportionately. Waterlogging occurs when the water level in field becomes so high that crop yields are substantially reduced. Improper farming and irrigation techniques following the abundant availability of water from the Lake Nasser have caused waterlogging and salinization of previously fertile land. However, extensive tile drainage is now being practised in Egypt to keep waterlogging and salinization under control. It is estimated that about 10 million hectares in India, or one quarter of the total irrigated area in 1978–9, are affected by waterlogging and salinization (Michael 1987). More than 70 per cent of 30 million hectares of irrigated land in Egypt,

Iran, Iraq and Pakistan are affected from these problems. It was estimated that over 2 million hectares of agricultural land were severely affected by waterlogging and salinization in Pakistan alone (Phantumvanit 1979).

As much as half the world's irrigation schemes suffer in some degree from waterlogging, salinization and alkalization. Globally waterlogging and salinization are reducing the fertility of some 1 million to 1.5 million hectares of fertile soil annually. Large areas being affected by these problems include the Helmud Valley in Afghanistan, the North Plain in China, the Punjab and Indus Valley in the Indian subcontinent, Mexcali Valley in Mexico, Imperial Valley in California, the Euphrates and Tigris basins in Syria and Iraq, the San Joaquin Valley in California, and Central Soviet Asia (Biswas 1978).

Increased irrigation efficiency, which conserves water, may still create problems (Water Quality Committee 1986). For example, with increased irrigation less water and therefore less salts in water will be added to the soil, but the absolute concentration (not the mass) of salts in the soil water of the root zone could be increased.

Seepage from irrigation channels is a common problem, and it often raises the water table in the area along the channel. A study of major irrigation schemes in Pakistan showed that seepage from unlined canals raised the water table by 7–9 metres within the first 10 years of operation (El-Hinnawi and Hashmi 1987).

Earthquakes. It has been proven that earthquakes can be induced by large reservoirs. The main earthquake in 1966 in Greece had its epicentre under the reservoir of the Kremasta Dam. Likewise, the 1967 Koyna Dam disaster in the Indian peninsula was caused by an earthquake whose epicentre was located at the dam itself (Biswas 1978). Extensive study during the filling of the Kariba Dam in Zimbabwe indicated that the earthquakes at Kremasta and Kariba dams did not occur by chance (ICOLD 1980). Seismic activity under the Kariba reservoir was strikingly parallel to the rise in water level over the five-year filling period, and the main earthquake occurred when water reached the top level for the first time. At Talbingo in Australia, only one minor earthquake was detected in the 13 years before the reservoir was filled, but 100 tremors were recorded during the first 15 months the reservoir was being filled. The strongest tremor occurred just when the water reached its top level. All the foci were shallow, just near the dam. Mild seismographic activities were persistently recorded in the basin after Lake Kariba had been built (Obeng 1978a). Similarly, minor seismic disturbances were experienced near the Sholiyar and Parambi Kulam Dams in India; this region did not have any recorded seismic disturbance before the project implementation (Varshney 1984).

Groundwater Mining. Excessive groundwater withdrawal may cause serious long-term environmental and social problems. Groundwater has been exploited extensively because of its simplicity and low cost. As explained above, apparent abundance of groundwater could be a result of accumulation over centuries and annual replenishment through rainfall could be minimal. Often groundwater cannot be considered as a renewable resource. The phenomenon of excessive groundwater pumping and subsequent lowering of the water table appears to be worsening around the world.

Effects on Climate. New microclimates are usually created by new large reservoirs. A change in the rainfall pattern has been observed around Lake Volta in Ghana, the peak having shifted from October to July/August. Evaporation from reservoirs in arid zones is very high, especially when the daytime temperature is high. During the night, if there is no wind, cool temperature thickens the vapour over the reservoir surface, creating fog banks. This fog remains on the reservoir parameter, causing some changes in temperature. The High Aswan Dam, by more than doubling the Lake Nasser's surface, has increased the total annual evaporation from 6 km to 10 km. The total annual loss due to evaporation is about 11 per cent of the lake water content. These changes of climate will ultimately affect humans and their activities such as cultivation.

Biological Impacts

Impacts on Living Organisms. The direct impact of dams and reservoirs would be the flooding of reservoir area, killing animals and plants in this area. Depending on local ecological conditions, dam/reservoir projects may be irreversibly destructive to wildlife. Unfortunately, wildlife management in many developing countries is still inadequate. It takes time and much resource to establish a full inventory of the wildlife in a specific region. Due to pressing schedule and lack of resources, baseline studies during the project planning stage often consist of superficial listing of the fauna and flora in the project area, with the result that species may be wiped out before their existence is ever known.

Furthermore, in developing countries direct destruction of natural flora and fauna may be only the beginning. Water development projects will open new roads and navigation channels into areas that so far have not been accessible. Coupled with the population increase due to migration from other areas, which increases cultivation and illegal hunting and logging, the wild ecosystem in and around the project area is in danger of destruction.

In dam/reservoir projects, much expectation has been placed on fishery in reservoirs. However, results may be disappointing without careful planning and design. For reservoirs to be stocked with fish, due to the combined effects of (i) stable stratification under tropical conditions as explained above, (ii) eutrophication, (iii) hydrogen sulphide toxicity, and (iv) siltation, the fish yield may be much less than expected. Experience seems to indicate that alterations in fish species composition are unavoidable. There is evidence that some fish species were lost and others reduced in number during the early period when the Lake Kainji was formed on the River Niger in Nigeria (Obeng 1978a). Surveys on the Nam Pong Reservoir in Thailand indicated that impoundment had severe impact on catfish but favoured murrels (Chatarupavanich 1979).

Ladders built to facilitate the movement of migratory fish to their breeding grounds do not always work. Where dams are more than 30 to 40 metres high, fish cannot move up the ladders without fatally delaying their migration.

There often occurs a reduction in fish production downstream of dams because the structures trap nutrients in their reservoirs. Downstream fishery was reduced by about 50 per cent after the construction of the Kainji Dam in Nigeria (Obeng 1978a). In Southeast Asia, natural fishery in the fields flooded by the river annual overflows constitutes an important low-cost protein source in the local people's diet. Thus, there may be profound and damaging effects of dam/reservoir projects on this type of fishery, but dam implementing agencies rarely consider mitigating this impact.

Damages on living organisms may extend far beyond the project area. Before the Aswan High Dam was built on the Nile, the river overflowed its banks every year between July and October as a result of summer rains in Ethiopia. The floods distributed silt, at a rate or 100–150 million tonnes per year, from the fertile basalt lava of Ethiopia over the Nile valley and delta. This supply of silt was one of the principal foundations of the civilization flourishing since early times along the Nile. The nutrients carried by floods also encouraged the growth of algae and fish off the Mediterranean coast as far away as Lebanon. Algae may be numbered as high as 2,400,000 cells/L during the flood season as compared with 35,000 cells/L during the dry season (Falkenmark and Lindh 1976). These algae provide food for, among other marine organisms, sardines whose catches were 18,000 tonnes a year. The Aswan High Dam traps 88–89 per cent of the silt as sediment in its reservoir, now releasing only 2 million tonnes of silt a year (Sundborg and Rapp 1986). The silt deposited yearly in Lake Nasser is equivalent to 13,000 tonnes of calcium nitrate fertilizer (Worthington 1977). By 1969, the sardine catches had fallen to a mere 500 tonnes a year. Meanwhile, the fish production in Lake Nasser also reduced when the dam was closed, then recovered to the normal level about 10 years later. However, it should be noted that as the fish yield in the lake continued its upward trend, the total fish catch in Egypt increased.

Increase in nutrient level in water may cause blooms of aquatic weeds, the most troublesome being the water hyacinth *Eihhornia crassipes*, the water lettuce *Pistia stratiotes*, and the water fern *Salvinia auriculata*. These weeds cause deoxygenation and eutrophication, impede transportation, increase water losses, harbour disease vectors, and suffocate aquatic organisms. The growth of aquatic weeds in newly formed reservoirs in hydroelectric projects in Africa is serious and spectacular (Obeng 1978a). In Surinam, aquatic weeds infested more than 50 per cent of the Brokopondo Reservoir within two years it was filled (Biswas 1978). Within three years after the Congo River was dammed, aquatic weeds spread throughout 1600 km of the river. The weeds absorb nutrients in water and when they die, the nutrients are locked up in sediments in the reservoir. Loss of water caused by weeds is due to two reasons. First, more water is required to compensate for reduced flow caused by the weeds. Second, owing to the large biomass weight per unit area and the high metabolic rate, the evapo-transpiration rate of aquatic weeds is very high. Once established, it is difficult, if not impossible, to eradicate the weeds. The Egyptian Government has spread massive quantities of herbicides—with unknown ecological consequences—on its canal systems, only with the hope of containing the weeds and reducing their effects to manageable proportions.

Impacts on Health: Water-related Diseases. Water resources development projects are closely linked to water-related diseases. These infections are caused by agents that have close relations with water, and are not new in the regions they exist. For example, schistosomiasis was known during the Pharaonic times as calcified schistosoma eggs were found in mummies dating back to 1250 BC, and the method (still practised) of removing the guinea worm from the patient's wound was described by Hippocrates. Under natural conditions, the agents of these diseases are contained by seasonal variations of water availability and by limited contact between humans and water. Water resources development projects promote year-round favourable habitat for these agents, increasing the prevalence of their infections and spreading them to areas never infested before. Year-round abundance of water also means increased contact between humans and parasites/vectors, and this not only increases case number but the severity of the disease (parasite load per patient). Large water resources development projects in Africa (such as Lake Kariba in Zambia/Zimbabwe, Lake Volta in Ghana, Lake Kainji in Nigeria, and Lake Nasser in Egypt) have almost invariably led to considerable increases in the incidence of water-based infections and those caused by water-related insect vectors (Hughes and Hunter 1971; Milligan and Thomas 1986; Obeng 1978b). In the northeast of Thailand, the liver-fluke *Opisthorchis viverrini*, which infects 60–80 per cent of the population in many provinces (Harinasuta 1975), has been of concern in water development projects in the region. In one survey, it was found that the rate of liver-fluke infection in irrigated areas of the Nam Pong Project (Ubolratana Dam) was 42.4 per cent, whereas in non-irrigated villages within the irrigation areas the infection rates varied from 8.8 to 18.6 per cent (Mekong Committee 1979).

Among diseases spread by water development projects, schistosomiasis (also called bilharziasis, after Dr Theodor Bilharz, a German physician who first identified the parasite) deserves attention. It is caused by three species of the flatworm genus *Schistosoma* whose life cycle depends on freshwater snails. People get infected when they come into contact with water bodies containing the parasites released from snails. They, in turn, spread the disease by contaminating water with their excreta or urine containing schistosoma eggs. The eggs hatch in water to release larvae which must, within 24 hours, find a suitable snail as an intermediate host, or they will die. The disease is of concern in water development projects in Sub-Sahara Africa, Arabian Peninsula, the Nile Valley, Iran, a part of the Middle East, Brazil, Venezuela, the lesser Antilles and Puerto Rico, China, Japan the Philippines and (in a small focus) Indochina. The number and its victims is currently estimated to be 200 million. After Lake Volta was formed, an explosive growth of the water lettuce *Pistia stratiotes* formed an ideal habitat for snails which are hosts of schistosome larvae. The schistosome parasites were present among people who had arrived from infested areas, and these parasites started infesting the local snail population. Soon, outbreaks of the disease were observed in the new townships along the lake. The rate of infection increased steadily, and within two years nearly all children were affected. In Egypt, a scheme for perennial irrigation of four areas led to an increase, within 3 years, of the percentage of the population infected with schistosomiasis in all four areas from 2–11 per cent to 44–75 per cent. The life expectancy of the men and women living in infested areas is 27 and 25 years, respectively. The estimates for the increase in schistosomiasis cases due to the Aswan High Dam are between 2.6 to 6 million.

Transmitted by anopheline mosquitoes, malaria, whose victims in the world number 200 million, is also often associated with water development projects. The Sarda Canal Project was started in 1920 in India with a plan to irrigate nearly 3 million hectares. The work was almost abandoned in 1929 as the labour force suffered heavily from malaria; in one team as much as 90 per cent succumbed (Waddy 1975). The process of clearing vegetation from water encouraged sunlight breeding mosquito vectors of malaria. The Gezira Irrigation Project with the help of the Sennar Dam irrigated nearly half a million of hectares of semi-arid savanna south of Khartoum, Sudan. At first the irrigated land was almost exclusively planted with cotton. This allowed simple but effective mosquito control by drying out irrigation ditches once a week—the time required for the life cycle of mosquitoes from eggs to adults. But gradually other crops were also grown, and the periodic drying out of the land was abandoned. Consequently, malaria begun to spread and infected more than half the population in some villages (Donaldson 1978).

Other major diseases associated with water development projects include:

- *Bancroftian filariasis*, which currently affects 250 million people. Transmitted by the bite of several species of mosquitoes, the parasite (*Wuchereria bancrofti*) lives in the lymphatics and causes the most characteristic symptom of elephantiasis, i.e. a massive overgrowth of the skin and subcutaneous tissues on the legs and scrotums, resulting in grotesque deformity and handicap.
- *African trypanosomiasis*, a fatal infection whose victims total nearly 35 million. It is caused by two parasites of the genus *Trypanosoma* and transmitted by the tse-tse flies which breed in fast-flowing water. The parasite invades the patient's central nervous system, causing disturbed cerebral functions such as sleepiness (the disease being called 'sleeping sickness').
- *Onchocerciasis*, which is caused by the worm *Onchocerca volvulus* and transmitted by the biting fly *Simulium* which breed in fast-flowing streams. The parasite itself normally causes no symptoms, but its larvae can enter the patient's eye, often leading to blindness. In Africa blindness is most commonly found near rivers in which the *Simulium* flies breed, hence the name 'river blindness'. The large dams in Africa have led to new exposures to onchocerciasis, especially during the construction stage as the velocity of the river is increased by narrowing the channel or by diverting it around the site. After construction (as in the case of the Volta Dam) breeding usually stops in the reservoir but may continue in the catchment area and spread the disease. For the Volta Dam Project, a 20-year international campaign involving several agencies and costing $120 million is underway to fight against river blindness, just one of the three major diseases spread by the dam project (Donaldson 1978). Although the programme led to an initial reduction of black fly populations, this reduction was achieved only temporarily. As the black fly is capable of wind-assisted dispersal over large areas, the area treated with insecticides is constantly invaded by flies from other areas.

When the project area attracts migration of people from other areas, these people may introduce new strains of locally endemic diseases or new diseases not normally found in the project area. On the other hand, newcomers may easily succumb to a disease that is indigenous in the project area. The eruption of malaria during the construction of the Panama Canal is a classical example.

An increase in water supply means an increase in resultant wastewater discharge. If the absorptive capacity of the soil is exceeded and without a wastewater collection system, open channels or pools of wastewater on the ground will serve as breeding sites for a number of diseases such as filariasis, haemorrhagic fever, intestinal parasites, etc.

Socio-Economic Impacts

Impacts on the Economy. Many of the impacts on the economy are derived from the impacts presented above. And they can even be quantified in monetary terms, such as the costs of reduction in fishery catches or of reduced soil fertility. Examples of impacts on the economy are:

- Loss of timber forest or low-lying farmland due to flooding, clearance, construction, or salinity erosion. Newly cleared land on higher grounds for compensation may not be as fertile as lost land.
- The deposition of silt upstream reduces the agricultural and/or fishery yields as well as natural biological productivity of the downstream areas.
- A development project attracts migration from other areas because it promises job opportunities or better basic services. Without long-term development planning, the project area will be burdened with an excessive population to feed, employ and house, leading to new socio-economic problems such as increase in living costs, unemployment, the development of slums, deterioration of public health, etc.

Without appropriate planning and management, potential benefits may not materialize. Fishery benefits of a dam/reservoir project may be substantial, but fish population management and suitable technology (methods and gear) used in catching fish are often ignored. Fishery on a deep and large reservoir is entirely different in many respects, from fishery on rivers that the local people are used to. Another problem in getting full economic benefit relates to distribution of benefits. Unequal distribution of a project's benefits has been a common problem in developing countries, when one group of people stand to benefit at the cost of another group. For instance, the total fish yield from a reservoir may increase but this yield is harvested by a small number of people. Furthermore, as has been experienced in Thailand (Harinasuta 1975), increase in fish production does not necessarily ensure increased income for fisherman, since most of the increased income may actually go to middlemen.

Impacts on Archaeological Sites. Before the Srinakarin Dam in Thailand was built, ecological surveys were conducted at the project area as part of an environmental impact assessment. The investigators found archaeological artifacts of the Hoabinhian age (named after Hoabinh Province in Vietnam where artifacts indicated a distinct prehistoric civilization age). The investigators (McGarry *et al.* 1972) recommended that this Hoabinhian area at the project area be excavated, any archaeological material found be moved to the National Museum, and during construction the workforce be alerted to the importance of the archaeological value of the area and report any sighting. Unfortunately, no significant follow-up was taken. The Thai Fine Arts Department was hampered by its vast area of responsibility and constrained by the lack of trained personnel. Consequently, the department was not able to render its full service to the investigation and excavation before construction (Phantumvanit 1979).

Before the Aswan Dam was built, the Abu-Simbell Temple located in the area to be flooded was successfully relocated to new site. This work, which was sponsored by Unesco, has been considered as a fine example of the preservation of archaeological treasures (ICOLD 1985). Another good example can be found in the Maduru Oya Project completed in 1985 in Sri Lanka. An ancient land and irrigation system was discovered under the surface when site clearing began. Archaeological investigations determined that some of the works were about 1600 years old, and of an extremely advanced technical quality. The Maduru Oya Dam was moved a distance upstream from its originally planned location to preserve the examples of early irrigation works.

Impacts on Socio-Cultural Structures. Water resources development projects may cause extensive upheaval on the socio-cultural structures of the project areas. The most well-known impact would be disruption of local people's livelihood in the area and their resettlement to a new area. In this regard, experience in developing countries has been a dismay. It is common to undertake a difficult and costly task of resettling tens of thousands of people for a single water resources development project. The Sanmen Gorge Project in China required relocation of 300,000 people who were unwilling to leave, and so created a number of problems (Wu 1986). The Pa Mong Project in Laos required resettlement of up to 460,000 people, or 16 per cent of the country's population. The total resettlement cost was estimated to be US$ 626 million (1975 value), according to Phantumvanit (1979). Approximately 100,000 people had to be relocated because of the Aswan High Dam.

Socio-economic impacts are complex and interrelated. Worse, many of them are intangible and easily cause controversies. Projections are difficult and good intentions may not give intended results due to peculiar and unpredictable human behaviour. In many African and Asian societies local people like to cling by all means to the land they are farming, to their deep-seated culture, their means of living, the food they are familiar with, etc. They will ignore all promises of a more civilized life. Local people, especially minorities, may have traditional social structures that outsiders (i.e. foreigners serving as project planners and designers) are entirely ignorant of.

In many cases, water development works are built in remote areas where ethnic minorities are not well understood or even acquainted. This was the situation in Quae Yai-Quae Noi Basin in Thailand, an ara of several major multipurpose projects and was inhabited by ethnic groups such as the Karen, Mon, and Lao (Phantumvanit 1979). It will be difficult to relocate people with their special cultural characteristics to a new environment. As to the people who have to leave their land, the experience is traumatic. When the Bennett Dam in Canada was built, the lifestyle of the local Indians and Metis was badly disrupted, and their social dislocation was serious. The situation was finally rectified at significant economic cost (Biswas 1980).

Quite often the work of resettlement has proceeded improperly, exacerbating the sufferings of relocated people. The Kariba Dam on the Zambesi displaced approximately 57,000 Tonga tribesmen, who suffered cultural shocks when moved to communities totally different from their own. An unsatisfactory settler/host relationship developed and led to a drawn-out conflict over land tenure. Being settled in a drier region, they had problems with planting and harvesting timings. They also faced severe food shortage as they could no longer catch fish and rodents (as they did when they lived on river banks) while land clearing did not progress as scheduled. And, ironically, they even suffered from water shortage (Obeng 1978a). But their hardship did not end there: in an effort to save the settlers from famine, the government had to send in food, and the food distribution centres became the transmission sites for trypanosomiasis (Milligan and Thomas 1986).

In another case, to build the Volta Dam in Ghana 78,000 people together with 973,000 animals in over 700 villages and towns had to be resettled. This was a nightmare since affected people previously living in small communities with different ethnic backgrounds, languages, traditions, religions, and social values and cultures, now had to live together in 52 locations (Biswas 1978). The complex emotional relationships between the different groups were not fully understood. Many of those being resettled near the newly formed reservoir were not fishermen, but had various occupations such as farming and trading, thus found themselves in entirely unknown hazards (Donaldson 1978). The impact of the change was terrible. Many people found it very hard to break altogether from their ancestral roots by leaving their gods, shrines and graves of ancestors. The adverse health that many relocated people experienced was not due solely to the new environment. Life disruption itself had harmful effects.

Despite the fact that socio-cultural impacts are complex in nature, they should be assessed to the extent possible. In the past, emphasis in project planning was biased towards physical planning. Gradually socio-cultural aspects were introduced into the planning process. International agencies, such as the World Bank, have insisted that the relocated people should be no worse off, and preferably better off, with the project. All resettlement costs, such as employment creation and compensation expenses, have to be incorporated in the project analysis. Detailed resettlement plans are essential before any project is to be appraised for assistance by the World Bank (Goodland 1986). While headquarters of agencies have a good intention, it may take time for local project personnel to seriously adopt the new policy direction.

CONCLUSION

This chapter is not meant to be negative towards development. It simply draws lessons from the past, so that what has happened will not be repeated in the future projects, and so that humans will have a good life.

All water resources development projects have environmental and social implications which are usually long-term, widespread and sometimes even irreversible. The real question is whether those changes are acceptable. It is not always easy to answer this question. One has to accept the hard fact that adverse impacts do happen, and very seriously. Perhaps the most well-known example of 'narrow-minded technology' is the Aswan High Dam. But still this does not mean that this dam should have never been built. It has doubled Egypt's electricity generating capacity, helped prevent disastrous floods, improved river navigation, created a vast potential fishery in the reservoir that has more than compensated for the sardine loss in the Mediterranean, and attracted more tourism. While the total construction cost of the dam (including subsidiary projects and electric power lines) was Egyptian Pounds 450 million, the annual return on full operation was 225 million (140 million from agricultural production, 100 million from hydropower, 10 million from flood protection, and 5 million from navigation). Dams and irrigation canals provide a lot of benefits, and more of them will be built. The question is how to avoid problems on our ecosystem which is so complex and still largely not understood.

While considering negative impacts, we should not ignore greater positive benefits, the real reason for development projects. Technology does exist to achieve the concept of balanced environment development. On the one hand, more water, food and other commodities have to be supplied to each person in an ever expanding population. On the other hand, the area of cultivable land is reducing and water is becoming scarcer each day. This necessitates a need for striking a balance between the environment and development.

REFERENCES AND BIBLIOGRAPHY

Biswas AK. 1978. 'Environmental implications of water development for developing countries', *Water Supply and Management*, vol. 2, no. 4, pp. 283– 300.

————. 1980. 'Water: A perspective on global issues and politics', in *Water management for arid lands in developing countries* (ed. AK Biswas), Pergamon Press, New York, NY.

Chatarupavanich V. 1979. 'Nam Pong "post-morten" environmental analysis,' Proceedings of National Seminar on Environmental Impact Statement— Guidelines for Water Resources Development, Office of the National Environment Board, Bangkok, Thailand.

Donaldson D. 1978. 'Health issues in developing country projects', in *Environmental impacts of international civil engineering projects and practices* (eds. CG Gunnerson, JM Kalbermatten), American Society of Civil Engineers, New York, NY.

El-Hinnawi E, Biswas AK. 1981. *Renewable sources of energy and the environment*, Tycooly International, Dublin.

El-Hinnawi E, Hashmi MZ (eds.). 1987. *The state of the environment*, Butterworths, London, UK.

Farid MA. 1975. 'The Aswan High Dam development project', in *Man-made lakes and human health* (eds. NF Stanley, MP Alpers), Academic Press, New York, NY.

*Falkenmark M, Lindh G. 1976. *Water for a starving world* (translated by RG Tanner), Westview Press, Colorado, USA.

Feachem RG. 1983. 'Infections related to water and excreta: The health dimension of the decade', in *Water supply and sanitation in developing countries* (ed. BJ Dangerfield), Institution of Water Engineers and Scientists, London, UK.

Gitonga JN. 1985. 'The occurrence of fluorides in water in Kenya and defluoridation technology', Report submitted by University of Nairobi, Kenya, to the International Development Research Centre, Ottawa, Canada.

Goodland R. 1986. 'Hydro and the environment: Evaluating the tradeoffs', in *International Water Power and Dam Construction*, vol. 38, no. 11, pp. 25–33.

Harinasuta C. 1975. 'Ubolratana Dam Complex, Thailand', in *Man-made lakes and human health* (eds. NF Stanley, MP Alpers), Academic Press, New YorK, NY.

Hughes CC, Hunter JM. 1971. 'The role of technological development in promoting disease in Africa', in *The careless technology—Ecology and international development* (eds. MT Farvar, JP Milton), Tom Stacey, London, UK.

Documents marked with an * are key reading materials.

*ICOLD. 1980. *Dams and the environment*, International Commission on Large Dams, Paris, France.

———. 1981. *Dam projects and environmental success*, International Commission on Large Dams, Paris, France.

———. 1985. *Dams and the environment—Notes on regional influences*, International Commission on Large Dams, Paris, France.

Jain RK, Urban LV, Stacey GS. 1981. *Environmental impact analysis: A new dimension in decision making*, Van Nostrand Reinhold, New York, NY.

Lee JA. 1985. *The environment, public health and human ecology—Considerations for economic development*, The Johns Hopkins University Press, Baltimore, USA.

McGarry MG, Amyot J, Arbhabhirirama A, Duangsawasdi M, Gorman C, Hudson N, Petersen S, Sabhasri S, Sidthimunka A, Sornmani S. 1972. *Ecological reconnaissance of the Quae Yai Hydroelectric Scheme*, Report prepared by the Asian Institute of Technology, Bangkok, Thailand, for the Electricity Generating Authority of Thailand.

*McJunkin FE. 1982. 'Water and human health', prepared for National Demonstration Health Project for US Agency for International Development, Washington, DC.

Mekong Committee. 1979. 'Nam Pong environmental management research project', Proceedings of National seminar on Environmental Impact Statement—Guidelines for Water Resources Development, Office of the National Environment Board, Bangkok, Thailand.

Michael AM. 1987. 'Environmental impact assessment of a major water development project in India: A case study of the Mahi Right Bank Canal system, Gujarat', in *Environmental impact assessment for developing countries* (eds. AK Biswas, Qu Geping), Tycooly International, London, UK.

*Milligan P, Thomas MP. 1986. Relationship between development and diseases, *Environmentalist*, vol. 6, no. 2, pp. 129–40.

Obeng L. 1978a. 'Environmental impacts of four African impoundments', in *Environmental impacts of internationaal civil engineering projects and practices* (eds. CG Gunnerson, JM Kalbermatten), American Society of Civil Engineers, New York, NY.

———. 1978b. 'Starvation or bilharzia?—A rural development dilemma', *Water Supply and Management*, vol. 2, no. 4, pp. 343–50.

Pearse PH, Bertrand F, MacLaren JW. 1985. *Currents of change. Final report: Inquiry on Federal Water Policy*, Ministry of the Environment, Ottawa, Canada.

Phantumvanit D. 1979. 'The environmental dimension of water resourses development', Proceedings of National Seminar on Environmental Impact Statement—Guidelines for Water Resources Development. Office of the National Environment Board, Bangkok, Thailand.

*Ramade F. 1981. *Ecology of natural resources* (translated by WJ Duffin), John Wiley, New York, NY.

Rau JG, Wooten DC. 1980. *Environmental impact analysis handbook.* McGraw- Hill. New York NY.

Roser SP. 1976. *Manual for environmental impact evaluation.* Prentice Hall, New Jersey, USA.

*Stumm W. 1986. 'Water, an endangered ecosystem', *Ambio*, vol. 15, no. 4, pp. 201–7.

*Sundborg A, Rapp A. 1986. 'Erosion and sedimentation by water: Problems and prospects', *Ambio*, vol. 15, no. 4, pp. 215–25.

Tuntawiroon N. 1985. 'The environmental impact of industralisation in Thailand', *The Ecologist*, vol. 15, no. 4.

Varshney DV. 1984. 'Case studies from India on the environmental impacts of dams', *International Water Power and Dam Construction*, vol. 36, no. 10, pp. 25–8.

*Verma RD. 1987. 'Environmental impacts of irrigation projects', *J. Irrigation and Drainage Engineering—American Society of Civil Engineers*, vol. 112, no. 4, pp. 322–3.

Waddy BB. 1975. 'Mosquitoes, malaria and man', in *Man-made lakes and human health* (eds. NF Stanley, MP Alpers), Academic Press, New York, NY. Water Quality Committee, Irrigation and Drainage Division. 1986. 'Report of Task Committee on Water Quality—Problems resulting from increasing irrigation efficiency', *J. Irrigation and Drainage Engineering— American Society of Civil Engineers*, vol. 111, no. 3, pp. 191–8.

*Worthington EB (ed.). 1977. *Arid land irrigation in developing countries— Environmental problems and effects*, Pergamon Press, New York, NY.

Wu Xiutao. 1986. 'Environmental impact of the Sanmen Gorge Project', *International Water Power and Dam Construction*, vol. 38, no. 11, pp. 23–4.

CHAPTER 2

Objectives and Concepts of Environmentally-sound Water Management

ASIT K. BISWAS

INTRODUCTION

Development in all fields during the past 3 to 4 decades has been significant, and water resources is no exception. For example, the number of dams throughout the world which have created large reservoirs having capacities higher than 100 million m^3, increased from a total of 1777 in 1951 to 2357 in 1985. During this period, major water projects were generally considered to be essential for further progress of developing countries in terms of simultaneously enhancing agricultural production and increasing hydroelectric power generation. This type of thinking was highlighted by Pandit Jawaharlal Nehru, the first Prime Minister of India, who termed the large dam at Bhakra as one of the 'temples of modern India'.

There is no doubt that for developing countries, which are all located in the tropical and sub-tropical regions with erratic rainfall patterns, efficient control and management of water has to be an essential requirement for their continued development. Without proper water management, self-sufficiency in food and energy will continue to be a mirage for most of these countries.

Scarcity of water and reliability of its supply are major constraints for agricultural development of arid and semi-arid countries. In most developing countries, many of all available sources of water which can be economically used have already been developed or are currently in the process of development. In some countries, such as Egypt or Jordan, there are already no new major sources of water that can be developed.

While the average amount of water available to each country remains constant, the demand for water is going up steadily for two important reasons. First, population in developing countries is increasing steadily. The world population in mid-1990 was estimated at 5.29 billion, which is nearly twice what it was only four decades ago in 1950. It is estimated that by the year 2000 the world population is likely to cross 6 billion. Currently, over three-quarters of the world population live in developing countries, and it is in these countries that 90 per cent of the projected population growth will occur. To provide the needs of this expanding population, more and more water will be required for domestic purposes, agriculture, industry, and hydropower generation.

Secondly, past experiences indicate that as the standard of living improves, the demand for water increases as well.

Thus, the major challenge facing water planners and managers in the 1990s is that while the physical availability of water in any country is fixed, its demand in all developing countries will continue to increase steadily in the foreseeable future. Accordingly, the problem is how to balance demand and supply of water under these difficult conditions since, unlike oil, water cannot be easily exported from a water-surplus country to a water-deficient country due to a combination of economic, political and environmental reasons. There is really only one solution, that is, to manage the available water resources in each country in an efficient and environmentally-sound manner.

ENVIRONMENTALLY-SOUND WATER MANAGEMENT

With the increasing awareness of environmental issues during the past two decades, there has been a growing concern about the adverse social and environmental impacts of water development projects in many developing countries. As discussed in Chapter 1, projects that have been intended to stimulate economic growth or provide more food have sometimes had adverse effects on agricultural lands, forests or fisheries, and on local communities. Doubts have been raised as to whether the goals that are being sought in water development programmes can be achieved without serious disruption of the environment or severe social impacts.

The matter of environment and development became an issue of increasing international concern in the mid-1960s. At that time the prevailing view seemed to be that environmental disruption was an inevitable price to be paid for economic progress. Since then, however, this position has been increasingly challenged in several fora organized by various United Nations bodies, in the scholarly literature and elsewhere. There is growing consensus today that, given certain preconditions, *both* economic development and environmental management can be pursued simultaneously (Tolba 1982). Arguments supporting this shift in viewpoint have focused predominantly on the water resources scene. It is here perhaps that the apparent conflict has been particularly sharp but also where the attempts to find an accommodation have been most sustained.

This chapter attempts to address the issue of the manner in which the philosophy underlying environmentally-sound water management can be encouraged and facilitated on an even wider scale. It takes the view that the key to reconciling the apparent conflict between development and environment lies in the incorporation of three critical elements in planning and policy making:

1. recognition of the concepts of sustainable development and resilience;
2. the adoption of a comprehensive viewpoint; and
3. the pursuit of higher levels of efficiency.

Any attempt to develop water resources results in some modification of the environment. Sometimes the impact is confirmed mainly in the river region, aquifer or lake itself, as in an alteration in the normal flow or the quality of the water body. In other instances the effects are much more widespread and may result in considerable alterations in land resources, forests or fisheries. Beyond this, water development may have major impacts on human settlements and economic activities. The seriousness of these impacts depends upon the ability of the various physical, natural and human systems to absorb them, as well as human perceptions about them (Biswas 1984).

To a significant extent the environmental impacts of water management are beneficial, particularly when they open up new avenues for economic development or social improvement without serious impairment of the resource base or the ecological system. The creation of a reservoir, for example, may provide additional habitat for the raising of fish or new opportunities for water-based recreation. Often, however, environmental consequences are adverse, varying in the degree of their intensity and social acceptability.

It should be clear from Chapter 1 that environment has a wide variety of meanings and that environmental disruption can take many forms. In some instances water projects may even result in the destruction of the resource on which they depend. It is also apparent that severe damage may be sustained by other resources or by activities in addition to those directly related to water development. At the same time experience across the world indicates that many of the problems of reconciling development and environment result from a failure to consider them simultaneously. Thus, in many countries there are ministries of industry and development which work in isolation from ministries of the environment. Often, too, the latter may be isolated from ministries dealing with various resources. Such separation clearly militates against the adoption of a holistic view. In addition, there is typically little attention given to possibilities for improving efficiency in the use of capital on the one hand and the use of water on the other. An overemphasis on construction alternatives has deflected attention from such options as demand management, which would not only reduce capital requirements but also be environmentally less disruptive. Various commissions and authors have observed that there is enormous waste in water supply systems, notably those involved in urban water supply and irrigation. They have also drawn attention to advantages of pricing mechanisms and which reflect the true costs of supply and the value of water in use.

Recognition of these disadvantages and deficiencies has led to increasing pressure for the specific inclusion of environmental effects as a consideration in water resources planning and policy making.

Environmentally-sound water management implies that:

1. development be controlled in such a way as to ensure that the resource itself is maintained and that adverse effects on other resources are considered and where possible ameliorated;
2. options for future development are not foreclosed; and
3. efficiency in water use and in the use of capital are key criteria in strategy selection.

Recognizing these ideas is one thing; translating them into action is another. What needs to be done beyond the preparation of handbooks, staging of training programmes and conducting of planning experiments? More specifically, what is required to foster the adoption of the three elements noted above in planning and policy making are: namely, the recognition of concepts of environmentally-sound development and resilience, the incorporation of a more comprehensive perspective and the pursuit of higher levels of efficiency.

None of the elements is entirely new, All have been introduced in varying degrees across the world. An examination of a number of themes of contemporary water management which incorporate one or more of the three elements will follow next.

THEMES OF CONTEMPORARY WATER MANAGEMENT

There have been some important changes in the approach to water management in various parts of the world during the past 30 years or so. Seven sets of concepts or themes may be identified, as follows:

1. the adoption of a comprehensive viewpoint;
2. the promotion of a search for the widest possible range of choice;
3. the recognition of water as an economic good;
4. the use of the river basin as a unit of area in various phases of water management;
5. the involvement of the public in planning and policy making;
6. the consideration of environmental impacts; and
7. the assessment of social impacts.

Comprehensive Viewpoint

The adoption of a comprehensive viewpoint implies the recognition of several important concepts in water management. One is that there are numerous potential uses of a given resource and that each has some effect on all the others. Occasionally, the pursuit of one may preclude several others, as in the use of a stream for the disposal of toxic wastes. Optimal use of the stream can only be attained if all the potential uses are considered simultaneously and a combination is chosen that conforms to a selected criterion, such as the maximum net economic return or the preservation of a particular ecological web.

A second concept is that use can sometimes be enhanced if several purposes are pursued simultaneously, that is, in a multiple purpose project. There are numerous possibilities, such as in hydroelectric power generation, flood control and irrigation development, as illustrated by the Columbia River scheme or in the development of the Tennessee Valley in the USA.

A third idea is that of the interrelationship of water development with the development of other resources. Water schemes often have effects upon land resources, as in erosion or the flooding of agricultural land. Similarly, land use practices can influence both runoff and water quality. Removal of vegetation or the introduction of irrigation are illustrations. Obviously the pursuit of particular goals in water development may impair the attainment of objectives in the management of other resources and vice versa. It makes sense, therefore, to identify such interrelationships and seek accommodation of effects in water planning.

Fourthly, a comprehensive viewpoint may imply the integration of water planning with overall economic and social planning. This is especially important when water is scarce and is vital to an economy, as in Egypt·or Jordan, or when water development is costly and may account for a large proportion of capital investment in a country. This is particularly true in developing countries; but it has become increasingly so in a number of industrially advanced ones as well, notably the United Kingdom.

Finally, the adoption of a comprehensive viewpoint has implied a recognition that water resources management requires inputs from a wide range of disciplines. Today the engineering profession is joined increasingly by specialists from such disciplines as economics, law, geography, sociology, biology and even psychology. A measure of the sophistication that water planning has reached may well be the number of disciplines that have been called upon to make concrete contributions to it.

The adoption of a more comprehensive viewpoint has been accomplished through a variety of modifications to existing institutions. These include the integration of *purposes* of management (such as provision of water supply, improvement of water quality, flood control, water-based recreation, etc.). Initially each purpose is undertaken independently, usually with separate legislation and administrative agencies. Eventually conflicts appear and economies of integration also become apparent. Larger, more comprehensive water agencies are then established. Increasingly today, the tendency is to link water development with other resource development, and, even more, to environmental management. The omnibus ministry of natural resources or ministry of the environment has become a widespread feature in North America, Western Europe and several other parts of the world.

A similar process of integration may be envisioned with respect to specific functions of water management, such as data collection,

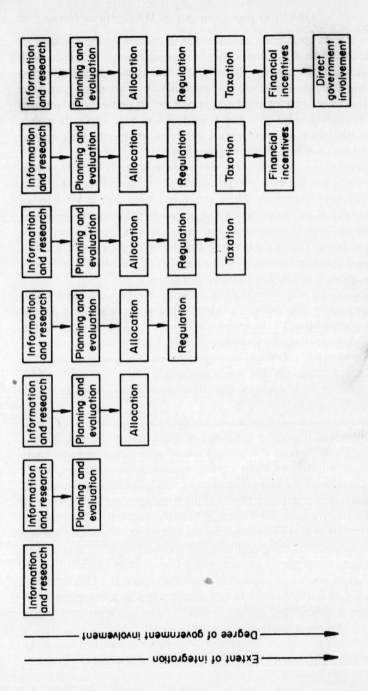

FIGURE 1. Integration of functions

research, regulation and development. As depicted in Fig. 1, there is a gradual increase in the scale and sophistication of functions. With respect to data, for example, the initial focus is upon the supply side; but it slowly expands to embrace the demand side and various economic, social and environmental considerations as well. It may be noticed that as the scale of management grows there is a growing tendency for government involvement to increase, reflecting in large measure the need to reconcile conflicts, provide public goods and to link water management to overall economic and social development.

Range of Choice

Theoretically there are several possible ways in which a given water problem might be dealt with or a water-related service provided. White (1964) and Smith and Handmer (1984) have shown, for example, that there may be as many as a dozen different strategies for reducing flood losses as shown in Table 1. An impending water shortage in a city might be accommodated in at least nine alternative ways (Table 2). Experience in North America, Europe and elsewhere has shown, however, that only a limited range of such options is typically considered by water managers and that this may lead to suboptimal choices, as reflected in higher economic costs, significant social losses or major environmental damages.

In several countries there is now a conscious attempt to canvass as wide a range of options as possible and to review the implications of adopting each of them. In the case of flood losses, for example, land use regulation and structural adjustments are considered together with various construction alternatives. Where water quality is at issue, incentives as well as penalties are now taken into account. Moreover, there are attempts to deal with the causes as well as the symptoms of pollution problems as shown in Table 3.

Water as an Economic Good

There has been mounting pressure from economists and others in recent years for water to be treated like any other natural resource, such as forests, fisheries or minerals: that is, to regard it as an economic good which can be traded in the market. Fees, it is argued, should be charged by the owner (generally the government) for its use and the price of services provided by its development should satisfactorily reflect the cost of supplying them. Marginal

TABLE 1. Classification of flood adjustments: structural/non-structural mode of implementation and the theoretical effect on losses

	Governmental[a]/Engineering[b]	Institutional[c]	Individual	Effect on losses
Structural	Dams Levees Diversions and channel improvements Retarding basins			Prevent losses
Non-structural		Acquisition Non-regulatory measures: fiscal and financial incentives, infrastructure provisions Regulations: zoning, sub-division and building regulations Information and education Forecasts, warning systems, emergency plans Salvage State and national emergency services Insurance Relief	House raising Small levees Other flood proofing local warning systems Response to warning Salvage	Modify losses Redistribute loss
Do nothing				Accept loss

[a] Government measures are those requiring a central authority to make and enforce regulations. to administer financial incentives/disincentives and to finance. construct and maintain major works. In the last case involvement may be through an agency regulating private enterprise.

[b] Engineering refers to the construction of major public works.

[c] Institutional measures are those requiring the direct involvement of government authorities through. for example. land use regulations. or their indirect involvement as guardians of the public interest. for instance by controlling the insurance industry.

TABLE 2. Alternatives for meeting increasing demands

Goals	Approaches	Techniques
Bear shortage or pollution		
Use available supply	Storage	Reservoirs
		Storm catchment
	Transport	Aqueducts
		Ground pumpage
		Motive transport
Increase overall supply	Precipitation inducement	Cloud seeding
	Increase and capture snow and ice-melt	
Improve water quality	Treat influent	Freshwater purification
		Desalination
	Treat effluent	Advanced waste treatment
Change water use	Reduce use or alter demands	Price curbs, metering
		Use restrictions
	Curb waste	Evapo-transpiration reduction
		Seepage reduction
	Alter distribution	Dual supply lines
		Direct pipelines
		Bottled water

TABLE 3. Opportunities for applying institutional instruments for water pollution control

Production–consumption stage	Policy instruments
Input mix	Use charges, taxes, pricing, raw material specification, subsidies/grants
Production	Land use zoning, taxes, subsidies/grants, process specification
Product output	Product specification, taxes
Product use	Taxes, pricing, subsidies/grants
Pollutant discharge	Collective treatment, (user charges) emission standards subsidies/grants, dilution rights/transferable discharge permits, withhold discharge during periods of low flow
Environment (ambient water)	Ambient standards, mixing zones specification

costs rather than average costs should be used as the basis for pricing such services. Economists also argue that water should be allocated among competing users or uses on the basis of the economic value derived, that is, the value in use. The higher the value, the more the user will be willing to pay. Finally, they suggest, the scope and scale of water projects should be determined through a comparison of benefits and costs. Such an evaluation would weigh the gains and losses of various scales of development and and various combinations of uses. Optimum scales and combinations would be those which produced the largest net benefits.

Recognition of such concepts has taken a variety of forms in water resources planning. Cost–benefit analysis and its more sophisticated derivations are now used by water agencies all over the world. There is an extensive literature describing its conceptual underpinnings and its potential applications. There are several handbooks that show how to use the technique.

In addition to introducing and refining cost–benefit analysis, economists have helped to improve water demand forecasting. At one time, estimates of future demands were based on little more than straight-line extrapolations of recent water use. No account was taken of such factors as the elasticity of demand or changes in technology or product mix. Research on such matters has materially helped improve forecasting techniques.

River Basin as an Unit of Area

The river basin has become increasingly accepted as a unit for water resources management. The rationale stems from the concept of a river as an organic entity, so that interference with or modification of any part of it will be felt elsewhere in the system. Engineers and geographers in particular have espoused the idea. Economists have recognized it, since it is an attractive way of internalizing externalities.

The extent to which the river basin has been adopted as a unit, however, has varied considerably. In England and Wales, France and India it is now used as the unit for all phases of water management in the country as a whole. In some countries it is employed for the management of water in specific regions, such as in the Tennessee Valley in the USA or in the Conservation Authorities in Ontario in Canada. Its most extensive use has been as a basis for the collection of data on water quantity and quality. Increasingly, however, it is being employed as the unit for water resources planning. Worldwide, examples can be found, as also manuals describing principles and setting out data requirements and procedures.

Public Involvement

Thirty years ago there was little attempt in most countries to consult the public on goals and strategies in water planning. The same was true in many of them even a decade ago. Generally, it was believed that planners and politicians could accurately gauge what the public wanted and how it would react to what was provided. Where public consultation did take place it was almost always at the end of the planning process. Mounting pressure, on grounds both of ethics and pragmatism, however, has resulted in a major shift towards a more direct role for the public in shaping options and in making decisions.

Public involvement, of course, can take a wide variety of forms as shown in Table 4. In some cases it amounts to little more than tokenism, with plans submitted for public approval, sometimes through the ballot box or in the public hearing at the end of the process. Sometimes there is no dialogue at all. Increasingly, however, a more productive role has been created in which a wider range of perspectives and talents can be called upon. The task force and the workshop are illustrative. Particular progress in this respect has been made in Canada, New Zealand and the USA. Attempts

TABLE 4. Strategies for public involvement

1. Public opinion polls and other surveys	
2. Referenda	
3. The ballot box	
4. Letters to editors or public officials	
5. Public hearing of inquiries	Increasing time commitment and value of feedback
6. Advocacy planning	
7. Representations of pressure groups	
8. Public meetings	
9. Workshops or seminars	
10. Protests and demonstrations	
11. Court actions	
12. Task forces	

are now being made in several developing countries, especially in Asia, to involve water users directly in water management process. In Pakistan it has now become a legal requirement.

Environmental Impacts

The assessment of the impacts of water development projects on the environment has come to be recognized as an integral part of planning. Until about 15 years ago there was little or no public concern about such matters and professionals in the water industry saw no reason to be worried about anything other than the manner in which environmental modification would affect the efficient operation of facilities. Today, however, water management agencies almost everywhere are highly conscious of the need to identify and evaluate the impact of projects on physical phenomena and on ecological systems. In some countries environmental impact assessment is now an elaborate process, often involving research and considerable interaction with various interest groups within the public. In some cases guidelines are provided as to what should be taken into account and how various impacts should be measured. Properly done, impact assessment involves a wide range of disciplines and a wide spectrum of interests (Biswas 1990; Biswas and Qu 1986).

Social Impacts

Over the past 30 years attention in the evaluation of water resources plans has gradually shifted from technical considerations to economic and environmental factors. More recently it has begun to embrace social impacts as well. Questions are being posed, for example, as to how far given schemes will disrupt traditional lifestyles, create ghost towns once construction is completed or set in motion a train of native land claims or a movement for a new political structure. Such questions have particular relevance in northern North America where clashes between quite different sets of cultural values are inevitable. Similar concerns have begun to arise in other areas too, notably in New Zealand where Maoris' lands and other rights might be involved, in Australia where there is growing pressure for the recognition of aboriginal rights, and in Narmada Valley in India.

Several responses have been made. One has been to highlight social effects in environmental impact assessments relating to water development projects. Another has been to foster the preparation of broadly based strategic plans for regions which are presently sparsely inhabited but which may come under considerable pressure for development in the future.

Each of the themes has resulted in a variety of institutional responses. Sometimes this has been expressed in wholesale restructuring of water management legislation, policies and administrative apparatus, as occurred in England and Wales in 1963 and in France in 1964. More often it has fostered institutional changes with respect to a particular water use or a particular water management purpose, such as reduction of damages from floods or droughts, as was the case in the USA, or Canada. A brief indication of some of the responses is provided in Table 5.

Experience in the adoption of the various concepts has varied considerably from one country to another. This has reflected the fact that problems vary both in their nature and severity from place to place. It has also reflected differences in political culture. In some countries there is an almost built-in penchant for change; in others there is an overpowering respect for the traditional way of doing things. Besides this, some countries are spurred into innovation by necessity while others can afford the luxury of biding their time.

It is possible to discern, therefore, a spectrum of adoption. Some countries have introduced several of the ideas with great enthusiasm. In contrast, there are a number where management today is just about the same as it was 30 years ago—or even longer. This is not necessarily bad. It may well be that the style adopted is appropriate to the physical, economic and cultural circumstances. Approaches tend to reflect stages of institutional evolution. One might expect, for example, that a country that has reached a high level of economic development, such as the USA, France or the UK, would tend to place high pressure on its water resources and would develop fairly sophisticated laws, policies and administrative arrangements to deal with the problems that have emerged. In many instances this would be likely to result in an increasing degree of government involvement. Such a situation might contrast vividly with that in certain Third World countries, such as Sierra Leone or Tanzania, where legal codes are simple, policies cover fewer issues and administrative arrangements tend to deal with single issues rather than a broad spectrum of problems in a coordinated manner. What is said about water institutions may also characterize institutions dealing with environmental management.

TABLE 5. Contemporary concepts in water management and associated institutional responses

Concepts	Institutional responses	Illustrations
1. Broadening perspective		
a) Coordinated water management	Omnibus national water agencies	India, Jordan, Israel, Hungary, Egypt, Sudan, Zambia, Zimbabwe
b) Links with environmental management	Ministers of environment with water resources branches	Canada, UK, France
c) Links with economic and social policy	Planning commissions or coordinating bodies	Israel, Hungary, France, India
d) Broadening range of professions	Professionals from disciplines beyond engineering, including law, economics, biology, geography	Canada, USA, UK
2. Expanding the range of choice	Policies to improve water use efficiency, such as recycling, wastewater renovation, planting of drought resistant crops	USA, UK, Israel, India, Egypt, Jordan
	Policies to supplement construction options such as flood insurance, land use control, flood plain mapping	USA, Canada, UK
3. Water as an economic good	Charges for resource use: water withdrawal charges, prices to reflect real costs	UK, France, Canada, New South Wales, South Australia, Zimbabwe, Israel
	Allocations to reflect values in use, metering	Fylde; UK; Denver, USA; Nairobi, Kenya
4. The river basin as a unit for management	River basin planning	USA, Canada, UK, France, India, Hungary
	River basin management	TVA (USA), RWA's (UK), ABF's (France), DVC (India), Gennossenschaften (FR Germany)
5. Public involvement	Ad hoc, usually at end of process, narrow range of methods	Most countries

TABLE 5. *Continued*

Concepts	Institutional responses	Illustrations
	Continuous, often	USA, Canada, UK, France required by law, using wide range of methods
6. Environmental protection as key element in water management	Specific legislation or clauses in water legislation	USA, Canada, UK, India
	Environmental impact assessment	USA, Canada, France, UK, New Zealand, FR Germany, India
7. Protection of minority rights and redress of social losses	Social impact assessment	Canada, New Zealand, USA, UK
	Settlement of native land claims	Canada, USA, New Zealand
	Institution of compensation measures	Canada, USA, New Zealand

CONSTRAINTS TO ENVIRONMENTALLY-SOUND MANAGEMENT

A voluminous literature exists at present which considers at least some aspects of environmentally-sound water development. In recent years it has generally been assumed, at least implicitly, that adequate knowledge is available on how to plan, design and manage water resources systems in order that environmental disruptions could be reduced to a minimum, and whatever residual disruptions occur would be considered acceptable to society as a whole. The real problem was thought to be not the lack of knowledge but appropriate application of the knowledge available to solve the problems. This 'application gap' was often considered to be the real problem, especially in most developing countries.

A comprehensive and critical analysis of existing literature on environmental aspects of water development indicates that there are many constraints which limit the potential application of available knowledge by water professionals and decision-makers in developing countries. On the basis of this analysis, the following four major constraints can be identified:

1. incomplete framework for analysis;
2. lack of appropriate methodology;
3. inadequacy of knowledge; and
4. institutional constraints.

It should be noted that the four major constraints identified are not independent. On the contrary, they are often closely interrelated.

Incomplete Framework for Analysis

The framework currently used for analysing and considering various environmental impacts associated with water development projects is overwhelmingly biased towards assessing only the negative impacts. Realistically, any reasonable water development project will have discernible impacts on rural development, environment and health, though the magnitude and extent of these impacts will vary from one project to another. Indeed, the very fact that any given project has been approved for implementation indicates that decision-makers expect it to have certain positive impacts on society; otherwise there would be no reason to waste scarce resources.

To a certain extent the emphasis on the negative impacts of projects and programmes is not difficult to explain. In the 1960s analyses of water development projects considered primarily technical and economic factors; environmental and social issues were generally ignored. Concerned with the adverse impacts of many development projects on society and the environment, a movement gradually developed to protect and preserve the environment. Environmental protection became an important political issue in the late 1960s, at least in many developed countries, through the activities of environmental pressure groups and non-governmental organizations.

This attitude to and perception of environmental protection was reflected in the United Nations Conference on the Human Environment held in Stockholm in 1972. An analysis of the Stockholm Action Plan that was finally approved by all the member countries of the United Nations would clearly indicate its negative approach to environmental management—stop all pollution stemming from any development activity, stop exhausting non-renewable resources, and stop using renewable resources faster than their replacement (Biswas and Biswas 1982). The emphasis was thus primarily on the adverse impacts of development.

Not surprisingly, environmental impact analysis, which was made mandatory in many developed countries in the late 1960s and early 1970s, was mainly concerned with the identification and amelioration of negative impacts; positive impacts were generally not considered. Because of this beginning, the term 'impact' has continued to have negative connotations.

So far as large-scale water development projects were concerned, another event of this period worth noting is the publication of a series of articles by Claire Sterling in the popular media on the adverse social and environmental impacts of the newly built Aswan High Dam in Egypt. She concentrated only on serious negative impacts of the Aswan High Dam, such as the loss of the Mediterranean fishery, an increase in schistosomiasis, salinity development, a reduction in the fertility of the Nile Valley through the absence of silt deposition, and coastal erosion of the Nile Delta. These articles, published at the peak of the environmental movement and in a 'small is beautiful' era, made the Aswan dam a *cause célèbre*. In retrospect such articles both helped and hindered later water development projects in terms of environmental issues.

They helped in the sense that social and environmental impacts of water development, which were generally a neglected subject around that time, became issues receiving due consideration. By drawing attention to these issues it was made clear to the engineering profession, which dominated and still dominates the water development field, that there are other important dimensions in addition to the techno-economic ones in which society is interested. Accordingly, increasingly more environmental and social impact analyses of water development projects have been carried out during the past decade. However, the emphasis has continued to be on the identification of only the negative impacts of water development and ways to ameliorate them.

Numerous examples can be provided for this all-pervasive bias. Only two examples will be cited here, one generic and the other case-specific.

On a conceptual basis, every time the health implications of irrigation projects are reviewed, the main consideration has been the presence of vector-borne diseases such as schistosomiasis and malaria. Irrespective of whether the increase in the prevalence of such water-borne diseases was due to water projects or not—an issue that will be discussed later—such an approach is not only biased but also somewhat simplistic and erroneous.

48 / ASIT K. BISWAS

Viewed in any fashion, irrigation is an integral part of rural development. As the project develops, agricultural production increases as well. With a better per capita food availability and a more diversified crop production, food and nutrition levels increase. The situation is further improved by increased livestock holding and development of inland fisheries in the reservoirs. An increase in the availability of animal protein is an important factor to consider in many irrigated agricultural projects, but has unfortunately been mostly neglected. For example, the mid-term evaluation of the Bhima Command Area Development Project in Maharashtra, India, clearly indicates the impact of increased livestock holding, even amongst landless labourers, on the nutritional status of the people (Biswas 1987).

In addition to improved food and nutrition, the health status of the rural people is further advanced by improvements in education, health facilities, the status of women, and general advances in the overall quality of life. This is shown diagrammatically in Fig. 2. At present, instead of considering the overall health situation in project areas, only the negative impacts are being accentuated.

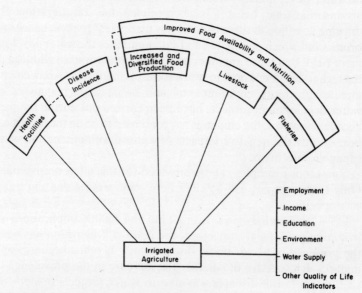

FIGURE 2. Pathways of interrelationships between irrigated agriculture and health

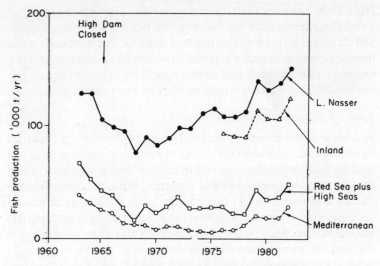

FIGURE 3. Total fish production, 1963–82

The case-specific example is the overall impact of the Aswan High Dam on fish production in Egypt. Starting from the time of Claire Sterling, much has been written on the decline of the fish catch in the Eastern Mediterranean due to the dam. On the basis of data available in FAO databanks, fish production in Egypt in the High Dam Lake, the Mediterranean and other areas was analysed for the period 1963–82. This is shown in Fig. 3, which indicates that fish production in the Mediterranean started to decline around 1963. The production of fish reached a minimum in 1975, but since then catches have been rising steadily. If the 'new' fish production system in the High Dam Lake is considered, total catches have always been significantly higher than at any time in the Mediterranean. Fish production started to decline in the High Dam Lake around the time of the closing of the dam, but it has now not only recovered but is higher than the initial production. The combined fish production in the Mediterranean and the High Dam Lake has been significantly higher than production from the Mediterranean alone before the Aswan Dam was built. Accordingly, the overall impact of the Aswan Dam on fish production in Egypt has been overwhelmingly positive, and not negative as most environmental literature suggests.

The beneficiaries, however, are not the same. Admittedly Mediterranean fishermen who did not wish to be relocated to the

High Dam Lake area have suffered serious economic hardships. What is thus needed is a balanced framework for analysis which will identify both positive and negative impacts. The next step should then be how to maximize the positive impacts and minimize the negative ones. A framework that considers only the negative impacts and ignores the positive ones is both incomplete and counterproductive.

Lack of Appropriate Methodology

A review of the processes currently used by developing countries to incorporate environmental issues in water management indicates that the methodologies available at present do not appear to satisfy the special requirements of those countries. While the environmental impact assessment (EIA) process was made mandatory in several industrialized countries, its actual use so far in developing countries has been somewhat slow. The reason for this slow acceptance is the lack of an operational methodology that can be successfully applied in developing countries with limited expertise, resources, data and time. The EIA methodologies that are being used in industrialized countries are not directly transferable to developing countries for various socio-economic and institutional reasons. Even in those developing countries where multilateral and bilateral aid agencies have carried out fairly comprehensive environmental impact analyses of water development projects, primarily with foreign experts and consultants, their overall effect in developing countries appears to have been generally minor. This is because such EIAs were carried out primarily to satisfy the internal requirements of the bilateral donor countries and multilateral funding agencies, and generally not at the behest of the developing countries in which the projects were located. Not surprisingly, the involvement and interest of developing countries in such external analyses have been minimal and somewhat superficial.

It is clear that complex, lengthy, expensive and time-consuming EIAs as practised in developed countries are not the right tool to assess the impacts of water development projects in developing countries. Under certain conditions complex EIAs may even prove to be detrimental, and they may hinder than enlighten the overall process of water development. As already pointed out, what is urgently necessary is the integration of EIAs into a broader multi-objective decision-making framework and the development of an operationally-sound methodology that is flexible and at the same time can be carried out within the limited costs, time frame and expertise available in developing countries.

Even though the United Nations agencies have put much effort into trying to develop such operational methodologies for developing countries, it has to be admitted that the process has been, for the most part, a failure. The guidelines prepared for various subjects are often elementary, and are of no use to any operational agency. Conceptually, as argued by UNEP's Regional Office in Bangkok, the EIA is only a 'pre-project' activity, which means that project impacts, even though analysed and identified properly, are not likely to change radically. Without any follow-up monitoring and implementation activities, the usefulness of EIA is reduced significantly and can at best be of marginal value. It becomes primarily a paper exercise to satisfy legal and institutional requirements and not a tool for effective impact management.

It is now clear that nearly all of what is now available in the area of EIA is of limited use for operational purposes in developing countries. What is necessary is to develop some guidelines which can actually be used by professionals for water management in planning and managing projects.

Lack of Adequate Knowledge

There are many areas where adequate technical knowledge may not exist for getting reliable answers. Equally, there are areas where 'conventional' knowledge can at best be dubious and at worst totally erroneous.

There are many areas where not only our knowledge is limited but we are also not even asking the right questions. For example, if the problem of vector-borne diseases is considered, probably the two most widespread and important are malaria and schistosomiasis. However, if we ask a simple question like to what extent a water development project *per se* may increase the incidence of malaria or schistosomiasis, there are no straightforward answers. The problem is further complicated by the case-specific nature of the answers.

In case of malaria, an exhaustive study by the Indian Malaria Research Centre indicates that the resurgence of the disease occurred independently of the green revolution. There is, however, no doubt that irrigation, agricultural practices, rice cultivation and migration of agricultural labour have all had an important bearing on mosquito vector fauna and malaria transmission (Sharma 1987). The linkages are not clear, and there is no evidence to indicate a one-to-one relationship between irrigation development and additional malarial incidences.

There are other complex issues that need to be considered for malaria. A study of two villages in the Kano plains of Kenya, one a newly established village within the 800-hectare Ahero rice irrigation scheme and another an older village nearby in a non-irrigated area with traditional mixed agriculture, showed remarkable differences in terms of differing mosquito species. In the new village 65 per cent of mosquito bites were from the *Anopheles gambiae* complex (principal vectors of malaria in tropical Africa), 28 per cent were of *Mansonia* species (vectors of lymphatic filariasis and Rift Valley fever), and 5 per cent were of *Culex quinquefasciatus* variety (another vector of lymphatic filariasis). In contrast, 99 per cent of the mosquitoes in the older village belonged to *Mansonia* species and less than 1 per cent *Anopheles gambiae*. Thus irrigation can change the transmission patterns of mosquito-borne diseases. This is an especially important consideration for tropical Africa where most of the global total of more than one million deaths due to malaria now occur (Biswas 1986).

There is also the issue of stratification. The evaluation of the Bhima command area development indicates that malaria appears to be attacking women more than men (Biswas 1987). How widespread this stratification is, either in India or elsewhere, is unknown since this type of question is not being asked at present, let alone being answered.

If schistosomiasis is considered, there is no doubt that the presence of an irrigation system in a developing country with extended shorelines of reservoirs and banks of canals contributes to a more favourable habitat for snails when compared to the pre-construction period. It will naturally have a tendency to increase the incidence of schistosomiasis. While no one would argue with this simple and acceptable fact, the question concerning the extent to which irrigation practices *per se* contribute to the increase in the incidence of schistosomiasis is more difficult, if not impossible, to answer at the present state of knowledge.

A perusal of the literature available will indicate a plethora of statements and so-called 'facts and figures' on the increase in schistosomiasis and other vector-borne diseases caused by the construction of irrigation systems. While it is accepted that such general statements had an important role to play in the late 1960s and the 1970s to sensitize engineers, decision-makers and general public to the importance of considering vector-borne diseases, very little progress has been made in the 1980s to give water planners and administrators the specific information they need to improve the planning and management processes.

One of the major problems with respect to the incidence of vector-borne diseases resulting from irrigation projects stems from the lack of an adequate number of scientifically rigorous studies. It is a subject that is replete with poor and conflicting information, repetition of data that have seldom been critically examined, and elucidation of personal biases. International organizations have to a certain extent contributed to this situation, albeit not deliberately. For example, the WHO's statement that globally some 200 million people are infected with schistosomiasis has remained remarkably constant since at least 1969. UNEP has incorrectly stated in the past that schistosomiasis has been completely eradicated in China. FAO publications have erroneously mentioned that water development significantly increases onchocerciasis, whereas all the evidence available indicates the opposite. The FAO (1987) has repeated examples of increases in schistosomiasis from water development projects based on poor and somewhat dubious data first published in 1978. A major problem in this area is uncritical acceptance and repetition of published information, irrespective of its quality. As these data get published time and time again, they gradually gain 'respectability'.

The second problem is the absence of data on pre-project conditions in terms of environment- and health-related factors. Even now, when some baseline surveys are being carried out on pre-project conditions, environmental and health issues receive virtually no attention. Without knowledge of pre-project conditions it is not possible to say with any degree of certainty whether vector-borne diseases have increased or decreased over time in a particular project area.

The third problem arises from the fact that objective and comprehensive evaluation of irrigation projects, including the incidence of vector-borne diseases, is never carried out at regular intervals. Accordingly, very little data exist on the basis of which realistic conclusions can be drawn. There are a few studies available on the incidence of vector-borne diseases in the project area, but they are seldom scientific or rigorous. Control samples are seldom taken. Equally little account is taken of the health status of people who migrate into the project area due to the expanding economic opportunities, even though some of the people migrating to a project area may already have been infected by vector-borne diseases.

In addition to these complex problems, there are three important issues that should be noted in any discussion of the implications of water development on the environment. First, the impacts of water development on environmental health are many. Some of these impacts are direct and comparatively easy to identify and to predict in advance. Others could be indirect and project-specific and thus often prove to be difficult to foresee and even more difficult to quantify. Most water resources projects produce a mixture of these two types of impacts. As is to be expected, it is less difficult as a general rule to predict and control primary impacts than secondary and tertiary impacts. Thus for impact analysis of any medium to large-sized irrigation project a substantial number of specific and interrelated factors have to be analysed, both concurrently and sequentially, in a coordinated manner within an overall framework, by a variety of professionals, based often on incomplete or unreliable data. Considering the methodological limitations that are inherent in such impact analysis, it is a difficult task under the best of circumstances.

Secondly, environmental impacts of projects, both direct and indirect, are never confined within the project boundary. Many of the impacts occur far from the project area. Accordingly, it is not possible to define a precise geographical boundary which could be said to contain all the impacts.

Thirdly, the time dimension of the impacts is another complicating factor. Certain impacts can be immediate, and thus can be identified during the implementation phase or soon thereafter. Other impacts, however, could be slow to develop, and thus may not be visible in the early stages. For example, some unanticipated changes in the ecosystem and the environment could take more than a decade of operation of a project before they begin to appear. For many impacts it is not possible to forecast the timing of their occurrence with any degree of reliability. A typical case is salinity development in irrigated areas, which could take 15–25 years in certain projects, but in others the problem may appear within 2–3 years, depending on physical conditions, drainage facilities, operation and maintenance procedures, and management practices. The time dimension also makes direct comparison of the impacts of different water development projects a difficult process.

Institutional Constraints

As pointed out earlier, a sectoral approach to water development is a major institutional constraint in all developed and developing countries, and this has an important bearing on the sustainability of projects. Medium to large-scale water development projects not only change the environment, but also have other substantial impacts on social well-being, among which are employment, education, health facilities, communication, energy availability, domestic water supply and the status of women. These impacts take place through a series of interconnected pathways which are both direct and indirect and are not always easy to identify or predict. They may also vary substantially in terms of both their nature and magnitude from one project to another. Unfortunately, a holistic approach to land and water management that considers all these issues is rare though attempts are now being made along these lines in a limited fashion in a few countries.

There are many reasons for this situation, but one of the most important is the division of responsibilities between the various ministries that look after various water-related issues. For example, the Ministry of Irrigation or Water Resources is responsible for irrigation development, the Ministry of Agriculture for agriculture-related issues, the Ministry of Health for health promotion, the Ministry of Environment for environmental matters, the Ministry of Education for schools, the Ministry of Rural Development for rural issues, and so on. Because of long-standing rivalries, the coordination and cooperation between the various ministries leave much to be desired. And yet in any large-scale water development project all these issues must be integrated within the project area. While it is easy to point out this necessity, how this integration can be effected in reality in the field is a very complex and daunting task. It has to be admitted that there are not many success stories.

CONSTRAINTS TO ACHIEVING
ENVIRONMENTALLY-SOUND WATER MANAGEMENT

There are many constraints which are inhibiting environmentally-sound water management in most developing countries. The importance of individual constraints could vary from one country to another, and sometimes even from project to project within the same country. Often the constraints are closely interrelated: one contributing to the other and vice versa. Among the main constraints are the following:

- Debts and financial deterioration in developing countries, which means lack of funds or substantial delays in allocating funds for essential requirements such as operation and maintenance of irrigation and drainage projects, deterioration in data collection activities, etc.
- Lack of appropriate and consistent policies for water development for both large- and small-scale projects.
- Serious delays in completing water projects after major investments like dams and other hydraulic structures and main secondary canals are completed. Significant delays in field canal construction and land levelling mean large areas which should be irrigated are not. Potential benefits are thus not being realized.
- Absence or inadequacy of monitoring, evaluation and feedbacks at both national and international levels.
- Lack of proper policies on cost recovery and water priciing or, if policies exist, absence of their implementation.
- Lack of professional and technical manpower and training facilities.
- Lack of beneficiary participation in planning, implementation and operation of projects.
- Lack of knowledge, and absence of appropriate research to develop new technologies and approaches, and absence of incentives to adopt them.
- General institutional weaknesses and lack of coordination between various ministries such as water, agriculture, environment, planning, etc.
- Lack of donor coordination resulting in differing approaches and methodologies, and thus conflicting advice.
- Inappropriate project development by donor agencies, e.g. irrigation development without drainage, supporting projects which should not have been supported.

CONCLUSION

Without environmentally-sound water management, it will not be possible for developing countries to achieve self-sufficiency in food and energy. During the past four decades, water development projects in terms of expansion of irrigation accounted for a major part of the increase in agricultural production necessary to meet the demand of an increasing population. Increases in agricultural yields resulting from introduction of irrigation was the main factor for enhancing total food production. Thus, by the mid-1980s, 36 per cent of the total global crop production came from less than 15 per cent of the land that was irrigated. This clearly indicates the importance of water development for the future survival and welfare of mankind.

The current global situation in terms of expansion of water control projects is a matter of some concern for a variety of reasons. For example, during the two decades between 1960 and 1980, the average rate of expansion of irrigated areas was 2.1 per cent. This rate declined by nearly 50 per cent during 1982 to 1987, when the average rate of expansion in developing countries was reduced to only 1.1 per cent. This compares with a population growth rate of 3.1 per cent per year in these countries. One important implication of this situation has to be that water must be used more efficiently than in the past if the standard of living in developing countries is to improve. Efficient use of water is simply not possible, unless it is managed in an environmentally-sound manner. Thus, environmentally-sound water management is an essential requirement for future development of developing countries, and will become an even more important consideration in the future than it was in the past.

REFERENCES

Biswas AK. 1984. 'Environmental consequences of water resources development', Key Note Address, 4th Congress, International Association for Hydraulic Research, Asia and Pacific Regional Division, Chiang Mai, Thailand, 11–13 September.

———. 1986. 'Land use in Africa', *Land Use Policy*, vol. 3, no. 4, pp. 269–85.

———. 1987. 'Monitoring and evaluation of irrigated agriculture: A case study of Bhima Project, India', *Food Policy*, vol. 12, no. 1, pp. 47–61.

Biswas AK, Qu Geping. 1986. *Environmental impact assessment for developing countries*, Cassell Tycooly, London.

Biswas MR, Biswas AK. 1982. 'Environment and development: A review of the past decade', *Third World Quarterly*, vol. 4, no. 2, pp. 479–91.

FAO. 1987. 'Consultation on agriculture in Africa', Irrigation and Drainage Division Paper 42, FAO, Rome.

Sharma VP. 1987. 'The Green Revolution in India and ecological succession of malaria', Paper presented to 7th Annual Meeting, WHO/FAO/UNEP Panel of Experts on Environmental Management of Vector Control, Rome, Italy, 7–11 September.

Smith DI, Handmer JW. 1984. 'Urban flooding in Australia: Policy development and implementation', *Disaster*, vol. 8, no. 2, pp. 105–17.

Tolba MK. 1982. *Development without destruction*, Cassell Tycooly, London.

White GF. 1964. 'Choice of adjustments to floods', Department of Geography Research Series 93, University of Chicago, Chicago.

CHAPTER 3

Planning and Decision-Making Framework

JANUSZ KINDLER

INTRODUCTION

The first two chapters of this book have been concerned with inter-actions between water systems and the environment, potential impacts of water projects, and the general principles of environmentally-sound water resources management. The emphasis has mainly been on interactions and principles. Although this is important for better understanding of water resources management, the use of this knowledge is ultimately necessary in one of the many possible situations when some water-related decision is to be made. This decision might be, for example, the choice of a water resources development strategy, physical location of an environmentally difficult water project, drought control or water use adjustment policy. This chapter deals with planning and decision-making in the area of water resources management, with special emphasis on random and unknown elements involved in these processes.

The usual structure of the planning process for water resources management is shown in Fig. 1. Whatever be the decision to be taken, one has to begin with the identification of the problem at hand and definition of the study objectives. Next, necessary input data must be collected, interpreted and processed for further analysis. In general three blocks of different data are required, describing: (1) water supply conditions (quantity, quality, variability); (2) water demand, water quality control and flood control tasks (quantity, quality, reliability); and (3) potential alternatives for balancing demand with supply, for water quality control and flood control (structural and non-structural measures). To proceed further, a good deal of thought must be given to the criteria which are to be used for evaluation of the planning or decision alternatives. Providing this is done, the proper analysis may begin. First, relatively simple methods are used to identify unfeasible for clearly dominatir g alternatives. Having somehow reduced the alternatives available to a limited number, one can then proceed to build and analyse much more accurate and closer-to-reality representations of the problem that needs to be investigated. The degree of sophistication of analytical methods applied at this stage will always depend on the available input data, the nature of a given problem, the character of the decision to be made and the type of expertise that may be available.

At this stage of analysis (Fig. 1) mathematical models of various complexity are often used. It was noted at the IAHS Symposium on Scientific Basis for Water Resources Management, held in 1985 in Jerusalem, that some of the most important deficiencies in modelling 'have been caused by an overenthusiasm to model, with insufficient attention and effort devoted to understanding the physical, economic and societal world which was to be captured in these models'. Moreover, model imperfections and limitations in coping with the random, unknown, imprecise and non-quantifiable factors should be recognized. This is why critical examination of modelling results is a very important consideration. Providing this is done, whatever decisions that need to be taken in a given context can be made with the assistance of the models.

In this chapter, mostly planning decisions, either of a long-term or a short-term nature, are discussed. The long-term decisions are usually of a strategic character—long lead-times are necessary for implementation of major water resources programmes. On the other hand, short-term decisions are of a tactical nature. However, as discussed elsewhere in this book, short-term decisions must be made within the long-term framework. Many times there is simply no time to wait for implementation of the long-term measures. Water authorities must react to changing situations by introducing new water use policies, adding some physical facilities to increase the reliability and capacity of the systems, changing operation rules of storage reservoirs, and other appropriate measures. In all types of planning efforts, assumptions concerning future management rules (operational aspects) should be made explicit and taken into account.

Since this book is concerned with large-scale water management, this chapter examines the notion of regional development. The development of a region means socio-economic and cultural moderni-

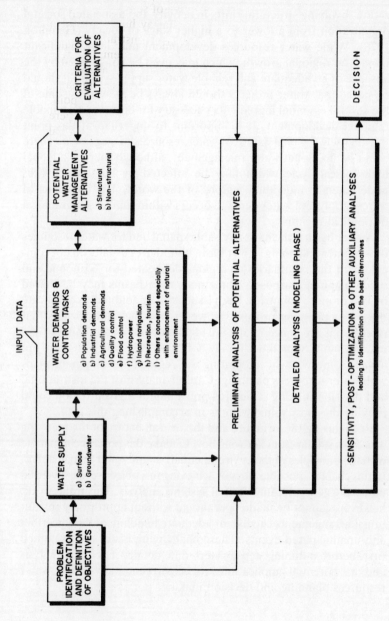

FIGURE 1. General scheme of water resources planning

INPUT DATA

PROBLEM IDENTIFICATION AND DEFINITION OF OBJECTIVES

WATER SUPPLY
a) Surface
b) Groundwater

WATER DEMANDS & CONTROL TASKS
a) Population demands
b) Industrial demands
c) Agricultural demands
d) Quality control
e) Flood control
f) Hydropower
g) Inland navigation
h) Recreation, tourism
i) Others concerned especially with enhancement of natural environment

POTENTIAL WATER MANAGEMENT ALTERNATIVES
a) Structural
b) Non-structural

CRITERIA FOR EVALUATION OF ALTERNATIVES

PRELIMINARY ANALYSIS OF POTENTIAL ALTERNATIVES

DETAILED ANALYSIS (MODELING PHASE)

SENSITIVITY, POST-OPTIMIZATION & OTHER AUXILIARY ANALYSES
leading to identification of the best alternatives

DECISION

zation involving structural shifts in activity and associated societal organization from a lower to a higher stage of society (Hamilton 1978). While water resources development may have significant impact on regional growth, which may even be constrained by the absence of an adequate and reliable water supply system, it should be noted that water projects should always be viewed as some of the several essential ingredients necessary for ensuring sustainable regional development. It is important to underline at this point the intensely regional nature of water resources management. There could be no global water management, although there is a global hydrological cycle which may be affected by human activities undertaken in individual regions of the world. The balancing of water supply and water demand occurs within smaller geographical or political regions.

A river basin is the most advisable spatial unit for water resources development planning. River basin boundaries usually do not coincide with the boundaries of regions delineated for economic and social planning purposes. Sometimes river basins may be shared by two or more countries. Such issues will undoubtedly complicate the water planning process. However, the fact still remains that river basins are conceptually the ideal units for water planning and management. Sometimes it may not be an easy task, but there are always some ways of integrating the regional development plans with water resources planning. Careful analysis of the institutional and decision-making structures prevailing in a given region could provide the most valuable hints in accomplishing this task.

Recognizing the structure and the overall nature of the planning process is still far from knowing how to make this process compatible with the principles of the environmentally-sound water management. In this chapter, possible ways of achieving this objective are reviewed under the general framework of systems analysis. This framework, however, cannot be made operational without appropriate institutional arrangements capable of adequate handling of the uncertain and unanticipated events. These challenging issues are examined first, before outlining step-by-step some details of systems analysis and its potential applications for environmentally-sound water resources planning and decision making.

THE INSTITUTIONAL ARRANGEMENTS

Although the importance of institutional arrangements for water resources management has been recognized by many, the term 'institution' is used in a variety of ways and for many different contexts. One of the most complete definitions of this term is given by Fox (1976), who explains that it refers

. . . either to an entity; an organization or an individual, or a rule, a law, regulation, or established custom. An institutional arrangement is defined as an interrelated set of entities and rules that serve to organize societies' activities so as to achieve social goals. Each nation has an institutional arrangement for managing water resources. This arrangement establishes the conditions under which water resources can be developed and used and provides organizations and individuals with certain resources and authorities to carry out the prescribed tasks. . . .

The institutional arrangements for water resources management have widely varying capabilities and limitations. Hence, these arrangements are not alternatives that can be compared in the abstract with a view to selecting the single 'best' institution. What will work best in a given basin depends not only upon its problem and its physical and economic characteristics but also on the general management milieu that is prevalent in the country. Further, in any particular situation, there should be an explicit need for a new institutional arrangement to accomplish specific objectives, before the best form of such an arrangement could properly be considered. If the existing agencies are carrying out their tasks successfully, one may consider how their performances can be further improved, but as a general rule it could be counterproductive to supersede or replace a reasonably functioning institution. Most often, however, major effort is required to devise and establish new institutions that could be strong enough to bring about unified and environmentally-sound development of the entire river basin.

Experience across the world indicates that most of the problems of reconciling water resources development and environmental quality result from a failure to consider them simultaneously in an integrative fashion from the very beginning of the project formulation process. Separation of the institutions that deal with water resources development and the environment, as well as inadequate coordination between the appropriate agencies concerned, clearly militate against the adoption of a more holistic view of water resources planning and management.

There are two principal trends in the institutional arrangements for water management, although they never appear in their pure form (O'Riordan 1985). One is towards centralization in an attempt to internalize the effects of decisions. The establishment of national water authorities responsible for managing both water quantity and water quality typifies such a shift. In such a case, public involvement and inputs originating from local initiatives tend to be reduced. Moreover, the information barrier typical for all large, centralized institutions could affect the overall quality of the decision-making process. Indeed, centralized water authorities are usually well informed about structural alternatives for water management such as storage reservoirs, interbasin water transfers, etc., but they often have much less information about the options available to individual water users for adjustment of their water demands. This deflects attention from the demand management alternatives which usually may reduce capital requirements and also could be environmentally less disruptive.

A second trend is towards decentralization, but in this case water authority must be equipped with some instructions to influence and coordinate the action of each and all independent water users. The extent to which water pricing schemes can be realistically used for this purpose is at least debatable, especially in developing countries. What other instruments that could efficiently be used is far from being clear. There is the problem of devising effecive market mechanisms which would conserve water and divert it from low- to high-value uses, where and when it may be socially desirable.

But centralization and decentralization should not be seen as two mutually exclusive options. Some of the most efficient management schemes feature strong national water authority collaborating with decentralized and, to a large extent, autonomous, regional (river basin) agencies. Catchment Authorities of New Zealand provide a good example in this respect (NZCAA 1987). They were set up in 1941 by the Soil Conservation and Rivers Control Act of Parliament to control flooding and erosion. In 1967, the Water Soil Conservation Act vested in them Regional Water Board responsibilities for the control of allocation, use and quality of water.

In New Zealand, catchment authority is a special purpose regional body which administers the water and land resources in one or more major catchments. A 'catchment' is an area of natural drainage by a river or stream, which contains a complex array of interlinked and interdependent resources and activities, irrespective of political or administrative boundaries. It is recognized that soil, water and vegetation resources cannot be managed in isolation from one another. In New Zealand, 20 catchment authorities have been established on this basis. There are four types of catchment authorities: Catchment Boards (13) which consist of a majority of directly elected representatives from the regions. Catchment Commissions (4) which consist of a majority of local representatives nominated by the constituent county, borough or municipal councils of the region, regional governments (2) of two major urban areas of New Zealand, and the Waikato Valley Authority in recognition for the hydroelectric development of the Waikato River. Work done at the regional level is overseen by the National Water and Soil Conservation Authority—the parent body of catchment authorities. The National Authority makes the policies that the catchment authorities carry out and allocates government funds for some work. It has 13 members from organizations such as the Catchment Authorities Association, the Municipal and Counties Associations, Federated Farmers, Manufacturers' Federation, and the Land Drainage and River Boards Association. However, to what extent the New Zealand experience can be successfully transferred to developing countries, or even to other developed countries, remains a subject of considerable conjecture. More about different institutional alternatives is discussed later.

It is important that institutional arrangements are capable of motivating all parties involved in water management decisions towards their effective implementation. There are situations where administrative structure seems to be ideal, laws and rules look perfect, but things do not work as expected. This often is caused by the lack of adequate motivating mechanisms. This difficulty has been most evident where authorities have relied chiefly upon regulatory process to achieve policy objectives. The environmentally-sound management of water resources should rely more on sequential decision-making, learning feedbacks and experimentation, that on detailed and usually inflexible policy schemes. Furthermore, for a variety of reasons, regulatory processes in most developing countries are not functioning properly.

To summarize, it is important that institutions charged with water resources planning and management are not too narrowly concerned only with the physical aspects of water management. They should be sensitive to changes in human preferences, habits, desires, aspirations and management abilities. The river basins, which are coherent hydrological units relevant to all water resources planning, design and operation, should be seen in such a context.

UNCERTAINTY AND RISK

Water resources are developed and managed under conditions of uncertainty, which arise from the underlying variability of geophysical processes. The natural stochasticity of these processes may be described by probability distributions, providing there is enough historical data. Furthermore, on the basis of recent experiences in many sub-Saharan countries, there is considerable scientific uncertainty as to the proper tasks of time series data that is necessary to define hydrological processes such as rainfall and runoff with some degree of reliability (Biswas 1988a). There are other sources of uncertainty which need to be considered because future is generally unpredictable with respect to economic events. For example, if the life period of a large dam like the Aswan is 500 years, there is a good possibility that accelerating human activities could change the hydrological regime on the basis of which a dam may have been constructed. The sources and degrees of uncertainty differ among various aspects of water resources planning and management programmes; they also differ in time. It is thus necessary to analyse various possible adjustments in these programmes, including evaluation of the related costs, to cope with the probable and the unforeseen events.

In water resources planning particularly important is uncertainty concerning future goals and the long-term availability of technical and non-technical means by which these goals could be achieved. These are highly uncertain factors external to water management for which the past offers little guidance. A retrospective examination shows that several large water projects have been built with little heed paid to the possibility that actual conditions might turn out to be substantially different from those assumed in the plans. There are no methods that could foretell the future, but by allowing for a more complete expression of uncertainty surrounding future goals and objectives, one can frequently point out more flexible, less risky and thus more sustainable alternatives. It should, however, be noted that water sector alone does not suffer from such uncertainties: all development programmes face similar uncertainties.

Many suggestions have been offered for tackling uncertainty, but in principle it may be dealt with in four different ways (Kaynor 1978). First, it may be ignored, but experience shows that planners can do it only at the peril of their plans. It may be possible to ignore uncertainties for small projects having limited life spans. However, to ignore uncertainty is to act arrogantly—as though the forecasts to take action were completely certain and anticipated and results were *faits accomplis*. This is contrary to any rational logic or theoretical structure.

The second option is to avoid uncertainty by incremental implementation of development plans. This may reduce the potential impacts of uncertain events, but it cannot eliminate all problems. Furthermore, this strategy, unfortunately, cannot be always followed in water resources development, especially in case of large-scale structural projects.

The third option is to reduce uncertainty, mostly at the research, data collection pre-planning stages. Better understanding of the cause–effect relationships is especially important in this respect. It should be noted that uncertainties may be reduced but cannot be completely eliminated. Furthermore, reduction of uncertainties generally has financial implications. Since the amount of funds available is limited, cost-effective solution that may carry acceptable levels of risks should be considered.

Finally, the fourth option is to view uncertainty as a chance occurrence and then incorporate the risk in the planning process. Risk analysis is playing an increasingly important role in assisting water resources planners to choose from different options related to the quality, safety and sustainability of the environment. It is generally a two-part endeavour. In the first stage of risk assessment usually a wide range of scientific disciplines is used to produce some estimate of risk.

Quite often, sensitivity analysis is employed in this stage, i.e. the technique of varying assumptions as to the behaviour of various factors. Next, those assumptions are examined in terms of the related economic and social benefits and costs. In the second stage of risk management, procedures and approaches to minimize risks are identified, evaluated and applied.

Although this may sound fairly straightforward, risk analysis, and especially the risk assessment phase, is a complex endeavour, often requiring a multi-disciplinary approach. The damage probability of a given event may be low, but if its perceived importance is considered to be high, it may require a high level of expenditures to reduce that perceived risk so that the project may be acceptable to the society as a whole. This is especially true when confidence people have on their knowledge of the consequences of that event is low (e.g. groundwater contamination by nuclear waste).

THE FRAMEWORK OF SYSTEMS ANALYSIS

Systems analysis, or the systems approach as it is commonly called, is aimed at considering a given entity in terms of the interrelationships between its components, rather than in terms of their individual properties. According to Miser and Quade (1985):

·A system analysis is an explicit analytic inquiry carried out to assist decision or policy-makers in a problem situation where uncertainty is present. The purpose is to help determine a preferred action or policy by identifying and examining alternatives and comparing them in terms of their consequences.

To avoid common confusion, modelling is not synonymous with systems analysis. To quote again the same authors:

A system analysis is an attempt to discern and answer questions of importance in the choice of a decision or policy; a model is merely a useful device in obtaining answers to such questions.

The use of mathematical models is not, then, all that systems analysis is about.

It should be recognized also that most, if not all, water resources issues and problems are characterized by multiple objectives, multiple criteria, multiple decision-makers, multiple users, and multiple constituencies. This often leads to the competitive or conflicting situations.

The methods of multi-objective analysis have been developed especially to deal with such competitive situations, when, for example, the improvement in one objective is associated with the deterioration in another. On a long-term perspective, however, economic, social and environmental objectives should not be viewed as competitive but rather complementary and mutually reinforcing. Accordingly, deterioration of the environmental quality may eventually hamper achievement of anticipated economic and social benefits that could arrive from any water development project.

The ultimate goal of water resources planning is to ensure that water of acceptable quality will be available in sufficient quantity at the right location, at the right time, at the right price, continuously on a reliable basis. Another developmental goal could be the protection against water excess—flood management and control. Because of the uncertainties inherently associated with all phases of water resources planning, all this can be achieved only within some limits of assurance, reliability and cost. Water resources development planning can vary significantly in scope, space and time. The following discussion, however, focuses on those features of environmentally-sound water resources development planning and management that emerge when large-scale water projects and long-term basin-wide strategic issues are being considered.

The overall approach of systems analysis is in principle always the same, irrespective of how advanced the stage of economic and social development of a given country may be. However, the level of sophistication that can be brought to bear on the analysis should be adjusted on the basis of expertise available, financial and time constraints, and the institution within which various alternatives will be considered. In general seven stages of systems analysis can be identified:

1. problem identification and formulation;
2. definition of objectives and their translation into evaluation criteria;
3. formulation and screening of policy or decision alternatives;
4. evaluation of alternatives including consideration of long-term environmental impacts and social effects;
5. selection of the 'best' alternative;
6. implementation of the selected alternative; and
7. ex-post performance analysis.

In most investigations iteration is needed among individual stages. The planner must be free to examine initial assumptions, to revise earlier work, and to impose additional constraints after analysing what some of the consequences of potential decisions could be. For example, there is always a feedback loop from evaluation to formulation of alternatives. As pointed out by the decision support approaches, it is also important to shorten the feedback loop between the planners and the decision-makers, with a frequent repetition of the study cycle. Thus, close interaction between the analysis and their clients is an important requirement for successful realization of the sustainable water resources development planning.

Taking these factors into account, the following discussion on how to incorporate environmental considerations into water resources planning is organized in accordance with the stages identified above. Explicit consideration of environmental concerns is necessary at each of these stages and various approaches may be used to accomplish this task. It should be recognized, however, that the planning process increases in rigidity from the problem identification to the implementation phase; thus insertion of environmental issues at the later stages becomes an increasingly more difficult and complex task. It also becomes an expensive process, and the results could often be sub-optimal (Biswas and Qu 1986). Raising environmental issues at the beginning of any planning process increases the chance that they will be considered appropriately during the entire planning and management phase and also is likely to generate more cost-effective solution than may be otherwise possible.

PROBLEM IDENTIFICATION AND FORMULATION

There is a general agreement that the appropriate problem identification and formulation is of a critical importance for the entire planning process. There is, however, a real danger that when there is a pressure to produce quick results, when the analytical resources and expertise are limited and the urgency of the resolution of the problems is critical, the importance of this initial stage of analysis may easily be overlooked. As stated by Quade (1980):

Problem formulation is concerned with such things as determining the goals to be achieved by a solution, setting boundaries on what is to be investigated, making assumptions about the context, identifying the largest groups, and selecting the initial approach the analysis is to take.

Planning for water resources development projects is initiated in response to needs that already exist as well as to needs that are anticipated for the future. Translating needs into a properly definable problem is itself a complex process requiring techno-economical skills and political-institutional insights. The problem becomes further complicated because the needs may vary with time due to changing conditions as well as changing attitudes and perceptions of the different groups of people involved. Initially, how the problem is formulated must be compatible with the language and precision requirements of those who are responsible for the initiation of the plan. A wide range of different situations may be encountered at this stage. Hence, problem formulation depends on the nature and scope of the problem, as well as the various technical, social, economic, institutional and environmental conditions within which the problem is to be resolved.

Among some of the main issues and constraints that could have a direct impact on the problem formulation are the following: (i) administrative and hydrological boundaries are seldom identical; (ii) time and budget allocations are always limited; (iii) various regulatory, legislative and political requirements are likely to narrow the range of feasible alternatives; (iv) water needs are predetermined exogenously to the planning process; and (v) adequate number of trained personnel may not be available. Furthermore, much may depend upon what may be perceived to be the main problem by the local population. For example, building a storage reservoir in a temperate area for a more affluent, nature- and conservation-oriented society, may be considered to be a serious environmental problem. However, construction for similar reservoir in a very arid or semi-arid developing country, where the reservoir is likely to contribute directly to much higher agricultural production and for hydropower generation, and thus contribute to poverty alleviation and employment generation, may not be considered to be a serious environmental problem.

The need for a clear and unambiguous problem formulation is critical, especially in developing countries, where data availability and reliability is a common and recurring problem and serious investigations must often be undertaken, solely to improve the information base.

Regarding identification of potential environmental concerns, checklists are particularly helpful at this stage of analysis. These checklists make use of impacts that might be expected from different activities and are based on past experiences. By itself this is valuable, but it says little about the magnitude, type, extent or the relative importance of the impacts. Nor does it give any idea about the time dimension of the occurrence of impacts. Various checklists currently exist (Biswas and Qu 1986). One example is that developed by US Agency for International Development (USAID) for assessing rural development projects in developing countries (USAID 1980).

This checklist contains several questions such as:

Will the project
- increase vector habitat?
- decrease vector habitat?
- provide opportunity for vector control?

There can be three answers depending on how much is known about the particular impact under consideration: Yes, No or Unknown. If an answer can be reached, then the checklist provides a qualitative classification for describing the intensity of the impact. There are eight intensity levels that can be considered: not determinable, high adverse, medium adverse, low adverse, low or insignificant, low benefit, medium benefit, and high benefit.

Many types of checklists are available. Among them are simple, descriptive, scaling, scaling-weighted, and questionnaire checklists. The potential uses of these checklists and their applications to water development projects have been discussed in detail by Biswas and Qu (1986). While checklists are useful and have certain advantages, they suffer from some limitations as well. They cannot assess the dynamics of change and the related uncertainties. Moreover, they deal only with the environment. Attention is focused only on the environmental impacts and project performance in terms of other objectives is not considered. Nonetheless, checklists, if used properly, are useful as initial guide, helping to ensure that important environmental factors are not left out of the analysis.

OBJECTIVES AND EVALUATION CRITERIA

An important step in the analysis is definition of the study objectives and translation of these objectives into quantifiable criteria. An objective is a goal that decision-makers seek to reach by means of the decision taken. Although the environmental objective is main focus of this book, it is just one of the several objectives of water development that must simultaneously be considered by the analysts. In the United States of America, for example, 'Principles and Standards for Planning Water and Related Land Resources' identify four objectives to be promoted through planning: (1) national economic development, (2) environmental quality, (3) regional development, and (4) social well-being. However, only the first two objectives are required to be optimized during the project planning process. The last two objectives are only 'accounts for displaying additional informaion, but not the principal factors in the final decision-making' (Eisel *et al*. 1982).

The choice of socially relevant objectives requires judgement both on the part of the water resources planner, and on the part of other participants in the planning process, especially the politicians. Planners need to get clear guidance on this from the politicians which unfortunately does not happen very often. This was stressed by Major (1977), who states that 'much of the confusion and debate about water resources projects that have been proposed in recent years has arisen because the planners were not developing design options responsive to the objectives of the political process'. This often occurred in the past because the water engineers who were responsible for the planning process considered it to be a technical issue, primarily in their domain, and as such to be decided by them. With changing social and environmental attitudes, this perception is changing as well.

Another reason has been due to the lack of understanding of the decision-making process by the planners and system analysts. Even though the existing literature on systems analysis for water management is replete with terms like 'decision-maker' or 'policy-maker', regrettably very few analysts have made any serious attempt to understand the decision-making process in the real world. As Biswas (1988) notes:

The 'decision-maker' referred to in nearly all papers published in the area of systems analysis is really a mythical person who has to decide between some alternatives conjured by the analysts, on the basis of some narrow criteria identified by the analysts that have very little bearing to the situation in the real world.

Fortunately some analysts have now realized this problem. They are making a determined attempt to understand the mindset of decision-makers and see first-hand how decisions are actually made. Such attempts should be welcomed and encouraged since they have the potential to significantly improve the present planning and decision-making process.

For comprehensive basin-wide planning studies, the objectives tend to be fairly global and they generally do not incorporate conflicting issues. They are intentionally made all-encompassing in order to ensure the broad support of the various constituencies and stakeholders. Negotiation and compromise are integral elements of the planning process, and for negotiations to succeed, the parties must start with acceptable agenda of project objectives that can be modified later. Having reached a consensus on general project objectives, more attention can be focused on translating these objectives into evaluation criteria.

An evaluation criterion is a rule used to measure the extent to which an objective has been achieved. Thus, the criteria used should be quantifiable. In long-term strategic analyses, there is a natural tendency towards limiting the number of evaluation criteria. Concerning the environmental objective, the use of an aggregated impact index has often been advocated. However, the aggregation process should not be carried too far. Each criterion should reflect, at least to some extent, specific aspects of environmental quality to be maintained and/or enhanced (e.g. water quality, land degradation, aesthetics, etc.). When the environmental consequences of a development alternative are presented in the form of a single numerical index derived by consecutive aggregation of different environmental quality measures, its validity and usability are mostly limited. For example, if various water quality characteristics are aggregated into a single water quality index, it may not give a clear picture of the situation except in a somewhat superficial sense. Use of such indices, under many conditions, could even be misleading.

Trying to merge too many non-commensurable entities into a single index value is usually an unproductive process, especially as the aggregation process often calls for arbitrary weightings and value judgements. The hypothesis that experts will be able to weigh numerically the consequences of, for example, one aspect of water pollution with another, or water pollution to impacts caused by upper catchment deforestation, is difficult to accept. On the contrary, there is a considerable body of evidence that experts do not like to answer such questions. Instead of considering an index arrived at by weighting schemes which can always be questioned and/or manipulated, the experts and decision-makers generally prefer to express their judgement on a set of explicitly stated evaluation criteria which describe in physical, economic and social terms the environmental consequences of a given development alternative (Miser and Quade 1985).

Although specification of evaluation criteria is always problem- and context-specific, it must be recognized that most of them cannot be defined by using the same units of measure and thus they are not intercomparable. Under these circumstances, it is inevitable that those responsible for the solution of the problem must become involved in the process of evaluating trade-offs among the degrees of achievement in terms of the various criteria.

FORMULATION AND SCREENING OF ALTERNATIVES

Identification of the proper course of action for water resources management requires careful consideration of all the feasible alternatives. As pointed out by Davis (1968):

. . . if the most desirable answers were generally evident, the solution would consist mainly of working out the technical details of a simple, straightforward engineering problem.

Except for very simple and perhaps small projects, this, however, is an exception rather than a rule. Any reasonable size multi-objective water development programme invariably will have a series of subproblems, each of which is likely to have a series of alternative solutions. Thus, a significant number of alternative solutions is always present in any reasonable size water project. The identification of the full range of potential alternative solutions is always an important and essential step.

The problem can usually be solved in a variety of ways, and the full range of choices should be explored. In developing countries, socio-cultural factors are of special importance in the process of formulating alternatives. Considerable evidence exists at present of water projects that have failed because the tradition and habits of the user community were ignored. Several examples are given by White and White (1978) and Biswas (1981, 1988), which clearly indicate the dangers of ignoring community choices and preferences by the planners.

In principle, three possibilities exist for formulating and screening alternatives (Haimes *et al.* 1987):

1. A small number of available alternatives could lead to the reduction or even eliminaton of the screening step;
2. A large number of available alternatives requires a preliminary analysis to explore the trade-offs between project objectives. This requires people with technical knowledge and experience in practical aspects of water resources development projects;
3. A large number of available alternatives requires the use of

hierarchical screening in stages, with an increasing rigidity of selection and/or exclusion criteria being adopted as the screening process proceeds.

For environmentally-sound water resources planning, special care needs to be taken to ensure that to the extent possible selected development alternatives do not foreclose other viable options. Where a choice is between preservation and development, and there are uncertainties with respect to future demands for the services of either alternative (e.g. recreation v. hydroelectric power development), the associated costs and risks should be taken into account. It should be recognized that technological progress generally increase the margins of substitution for water in the production of commodity goods whereas it is generally incapable of augmenting the supply of environmental resources.

EVALUATION AND ALTERNATIVES

After the development alternatives have been reduced in number through the screening process to a few select ones, it is necessary to have some means of predicting, and whenever possible measuring, the environmental impacts of each development plan. It should be noted that all large-scale water resources development projects will contribute to both positive and negative impacts, and that the analysis should not concentrate exclusively on the negative impacts— as is often the case at present (Biswas 1988b). Current methods of estimating environmental impacts include Environmental Impact Assessments (EIA) and simulation models of various complexities.

Most EIA methods are based on matrices or flow diagrams which are designed to ensure that all potential interactions and impacts are at least included, with some indication of their relative importance. Among matrix methods, the earliest one used is known as the Leopold Matrix (Leopold *et al.* 1971), which consists of a horizontal list of development activities ranged against a vertical list of environmental criteria and conditions. Within each cell, the magnitude and importance of each possible impact are ranked on the scale 1 to 10. Several other types of matrices have now been developed, some of which have been used successfully in analysing the environmental impacts of water development projects (Biswas and Qu 1986). Flow diagrams which illustrate cause/effect relationships are also available, but the consequences of a variation in the design of a water project can be taken into account only by constructing another diagram.

A major criticism of all these approaches is that they are somewhat mechanistic, and provide only a set of static pictures of reality. They take little account of the interrelationships between the different environmental processes and the combined effects they can produce.

Simulation models are designed to track explicitly the dynamics of systems. A number of large-scale ecosystem models have been developed during the last two decades. Their development was largely stimulated by the series of biome studies (grassland, desert, tundra, deciduous and coniferous forests) carried out within the framework of the International Biological Programme (Patten 1975). These biome models have been useful in conferring better understanding of ecosystem structure and dynamics, but they do not readily lend themselves to answering specific questions concerning resource management. Many of those models are mathematically quite sophisticated. They contain provisions for non-linear relationships and can represent fairly complex environmental processes. But at the same time they often ignore those aspects of the ecosystem behaviour which are of major interest to water resources planners. Their actual use in planning and management of water development projects is, for all practical purposes, negligible.

There are also several simulation models which concentrate on hydrological or water quality impacts. For example, the well-known Stanford Watershed Model which, among other purposes, can be used to simulate sediment transport processes and to evaluate land use changes in river catchments. Another model, QUAL2E can simulate the behaviour of up to 15 water quality constituents in the river system in any combination desired by the user (Brown and Barnwell 1986). However, data requirements for these models are demanding and that is why they have not been often applied in developing countries. Their use in developed countries has been limited.

Several groundwater models have also been developed, which allow for the simulation of flow dynamics and dispersion of contaminants in aquifers. Although many of these models are now readily available and easy to run on relatively inexpensive microcomputers, they again require considerable input data and technical expertise in data acquisition. This presents a major problem for the application of these models in many developing countries. Furthermore, it has to be remembered that present knowledge of many environmental processes and cause-and-effect relationships is still very poor and the use of models should never give the impression that all is known and that all impacts as well as their magnitudes can be predicted with adequate reliability. Indeed, there is a considerable uncertainty associated with predictions made by these models.

The experience of the United States Water Resources Council is relevant here. During the years 1973–82, special effort was made by the Council to improve specific procedure that could be applied uniformly and consistently to evaluate appropriate beneficial and adverse effects of planning alternatives. Although significant progress was achieved, the Council still considers imperfections of such procedures to be a major obstacle to more complete realization of the multi-objective intent of the official 'Principles and Standards for Planning Water and Related Land Resources'. This problem is considered to be more fundamental than the difficulties related to analytical complexities of the multi-objective methods and techniques (Eisel *et al.* 1982).

CHOICE OF THE 'BEST' ALTERNATIVES

A whole range of quantitative and qualitative methods is being offered by systems engineering, operations research and management science for ultimate selection of the most promising planning alternatives. The focus of this selection is on the multi-criteria methods of analysis. Although the use of 'objective' to mean 'criterion' is incorrect, the terms 'multi-objective methods' and 'multi-criteria methods' have practically the same meaning. During the past two decades, these methods have experienced spectacular growth, capturing the attention of many and bringing about some new theoretical developments in the field of water resources management (e.g. Haimes *et al.* 1975; Cohon 1978; Chankong and Haimes 1983). Although one would be hard pressed to identify even one successful case where multi-criteria method has been actually used throughout the entire course of planning, even in a single developed country, they do appear to provide a viable framework of analysis for environmentally-sound and sustainable water resources development and management.

There are several reasons for relatively slow adoption of the multi-criteria methods (Biswas 1988b). Some of them require information which is difficult to obtain in a real-world setting. Sometimes the complexity of mathematical formulations puts off the potential users.

Generally, prevailing institutional arrangements are not favourable for their application. But as Cohon (1978) has stated: 'real world problems are multi-objective (multi-criteria) and imposition of a single-objective approach on such problems is overly restrictive and unrealistic'. There may be problems with practical introduction of certain methods or techniques, and/or with the abilities of existing institutions to carry out particular studies, but not, at least conceptually, with the multi-criteria framework of analysis as such.

The multi-objective methods of analysis represent a generalization of the single-objective approach—instead of a single objective to be optimized, multi-objective analysis deals with several objectives simultaneously. For such problems the concept of optimality must be dropped because a solution which maximizes one objective will not, in general, maximize any of the other objectives. This leads to the notion of a non-inferior solution (sometimes also called non-dominated, poly-optimal, efficient or Pareto-optimal), where any improvement in one objective can be achieved only at the expense of degrading another (for example, improvement of short-term economic gains at the expense of natural resource degradation). A set of non-inferior solutions sometimes appears in the literature under the name of a transformation function.

The problems that yield to multi-objective analysis may be classified in many different ways, although distinction is usually made between continuous and discrete problems. Continuous problems encompass an infinite number of choice possibilities that meet the constraints of a system under study, while discrete problems display a finite number of feasible choices (courses of action, strategies, solutions, project alternatives, etc.).

For a better understanding of the basic concept, let it be assumed that there are five feasible project alternatives in a three-criteria problem of economic efficiency, water quality and land degradation are given as in the following table (adapted from Cohon 1978):

	Alternatives				
Criteria	A	B	C	D	E
Economic efficiency	5	4	4	3	2
Water quality	8	9	4	10	9
Land degradation	7	2	4	6	8

Let it be further assumed that the numbers given in the table represent rating of each criterion on the specific project alternative. The more efficient or better in terms of a given criterion is the alternative, the higher is the rating. If we consider a single criterion of economic efficiency, alternative A is certainly the best (optimal) one. Alternatives D and E are optimal for water quality and land degradation respectively. Under these conditions, what can be done when all three criteria have to be simultaneously taken into account? First, the notion of non-inferiority must be defined. To determine the non-inferiority of an alternative, its ratings for all three criteria are compared with the ratings for all other alternatives.

Consider first alternative A. Projects B, D and E give better water quality, but are less attractive from the point of view of economic efficiency and land degradation—project C gives less in terms of all three criteria—project E yields better water and less land degradation, but is worse in terms of economic efficiency. Therefore, alternative A is non-inferior. Comparing alternatives with one another leads to the conclusion that only project C is inferior (is dominated by project A) and projects A, B, D, and E are non-inferior.

But even in a multi-criteria analysis, it may happen that there are some other criteria important to the decision-makers that have never been explicitly articulated. In such cases, the 'best' alternative may be located among the inferior alternatives and not necessarily among the non-inferior ones. These are only the words of caution that rejecting automatically all inferior alternatives is not advisable.

An additional limitation of the analysis to the non-inferior set of alternatives is that they are not comparable. For example, alternative D is better than E in terms of economic efficiency and water quality but it is worse for land degradation. Which one is better? Is it worth giving up one unit of water quality and one unit of economic efficiency to gain two units of land quality in moving from project D to E? This obviously involves the choice between competitive criteria and the degrees to which each of them is satisfied.

Selection of the most preferred or so-called 'best compromise' alternative requires additional information on the trade-offs between achievement of different criteria to be identified and evaluated. This evaluation is never a purely mathematical or quantifiable exercise. To varying degrees, it reflects the economic, social, aesthetic, cultural or even moral values accorded to individual criteria by those who are charged with the decision-making responsibilities.

This is undoubtedly the most difficult stage of the analysis and several methods have been proposed as aids in the search for the best compromise alternative. These aids can be found in any of the numerous textbooks on multi-objective (multi-criteria) analysis and planning. Since these methods allow for non-commensurate criteria to be traded without artificially combining them, it is a significant step forward in analytical capability.

IMPLEMENTATION OF THE SELECTED ALTERNATIVE

Major projects in developing countries usually call for step-by-step implementation, rather than immediate concerted action. Due to the general shortage of data and the limited understanding of all possible environmental impacts, continued monitoring of impacts should be an important part of the implementation process.

It must be recognized that uncertainties are always present and better results can be achieved if the impacts of development are monitored from the beginning of project implementation. Only then can an unexpected impact be identified early on and the plan or policy appropriately modified. This is an especially important consideration since at the current state of knowledge it is not possible to predict all environmental impacts, the time when they may occur, or their magnitudes. Only monitoring can identify problems as they arise, and steps can then be taken to ameliorate such problems. Continuing monitoring of water development projects is a prerequisite for environmentally-sound management, and yet this has for the most part been neglected so far in most countries.

PERFORMANCE ANALYSIS

As soon as the project or policy becomes operational, one of the issues of critical importance is analysis of performance, aimed at determining the extent to which the objectives of the project or a policy are being achieved. Such analysis may, for example, ascertain how much water from an irrigation project is actually used, what the originally unanticipated responses of project users are, or what the impact of the project on the downstream groundwater regime is. Such a performance analysis is indispensable for assessing the effectiveness of the project or policy initiatives and it may help prevent mistakes in the future.

Unfortunately, there tends to be a reluctance to examine past experience, simply because of a fear that objectives have not been attained or that unanticipated effects have appeared. While such reluctance is understandable, especially when analysts may share some of the responsibility for the past decisions, the future is better served by a system that creates positive incentives for the pursuit of this kind of feedback in operation. Competent evaluation of the performance of the water development projects in contrast to pseudo-evaluation also is a problem that seems to be pervasive in many bilateral and multilateral aid agencies (Biswas 1988).

SYSTEMS ANALYSIS IN DEVELOPING COUNTRIES

As already mentioned in the preceding sections, systems analysis can and should be applied in the developing countries, although the extent and means of its application may vary from one country to another. Making general statements about these differences is difficult, for what is valid for one developing country may not be applicable at all to another, as these countries are not homogeneous and could be at different stages of development. But at a risk of some oversimplification one could say that in preparing water resources development plans in developing countries emphasis should always be on the development of people and not on material things. If development is to benefit the people, as discussed in detail elsewhere in this book, local communities should be encouraged to participate throughout the project cycle, i.e. from planning to the operation and maintenance stage. This is the best way to create the sense of ownership and to provide for the early transition of responsibility from the government to local communities for the operation and maintenance of their water projects.

The socio-cultural factors are of particular importance in all stages of analysis, especially in formulating planning and policy alternatives. Many problems have emerged in developing countries due to the lack of understanding of the local conditions and value systems. Uncritical importation of foreign technology is a classical example in this respect. In many instances it has contributed not only to technical problems but also to the increase of social disparities as well. Alternatives and solutions should come first from observing the local environment and carefully examining the traditional strategies, and next from outside inspiration.

While the use of systems analysis is increasing, its potential for improving water resources planning and decision making has not been fully explored in the developing countries. Among major factors considered responsible for this state of affairs are: shortage of expertise and trained personnel, shortage of reliable data, and lack of confidence in modelling, which although not synonymous with systems analysis is one of its main tools.

As pointed out by Biswas (1981), the first factor is easily overcome since people can be trained quickly—'provided those who are in power are willing to see this done'. Data availability in developing countries is certainly a more acute problem than in developed countries because of the generally much shorter period for which measurements, observations, and scientific studies have been in progress. But it should be recognized that systems analysis can assist in data collection and processing. Moreover, the analysis can and should always be adapted to the limitations of data availability.

The lack of confidence in modelling cannot be considered to be a special attribute of developing countries since, to some extent, it may be observed all over the world. Admittedly, models are often built seeking academic rather than policy goals, but fitting the problem to familiar mathematical tools rather than tools to the problem (this has been called by Biswas [1981] solution-in-search-of-a-problem approach), or by describing what one can describe, not what is relevant (Miser and Quade 1988). But this is not to diminish the value of modelling, which if properly carried out can be of great assistance, even when insufficient quantitative data or only qualitative information is available as input to the model. A possible way out of such difficulty suggested by Frenkiel and Goodall (1982) is to build

. . . a simplified model based on previous knowledge regarding the identification of the relevant variables, and of the existing relationships between the variables; this simple model could then be used to determine at least the qualitative effects which relevant variables will have on the environment. The results of such studies may become useful to decision-makers in understanding the potentiality of simulation modelling, and a more advanced simulation modelling study with the necessary quantitative data base could then be initiated.

It should be recognized that problem-solving is not the only payoff from systems analysis. In many instances, its most important contribution is simply to increase understanding of some relevant concerns or to force rethinking on a complex problem. In this regard systems analysis may also be able to derive substantial benefits in the developing countries.

In the past decade, water resources planning has relied increasingly on concepts and procedures of systems analysis. It should be recognized, however, that water resources planning possesses a number of difficult-to-deal characteristics, the most important of which are uncertainty and underlying diversity of problem perceptions, objectives, criteria, and constraints.

As pointed out by Liebman (1976) more than a decade ago, single-objective optimization cannot and should not be used to resolve conflicts due to this diversity. It can be very useful, however, in illuminating these conflicts and generating non-inferior alternatives for further exploration. Using optimization in this fashion would save a lot of disappointment in some of the past applications of systems analysis to water resources planning.

Few observations that seems to be particularly important for environmentally-sound water resources planning and decision-making are the following:

1. Water resources planning should be linked on a continuing basis with the social and economic institutions of each country, as they also change and develop.
2. Appropriate treatment of uncertain, unknown and risky factors is imperative.
3. More emphasis should be laid on understanding decision-making environments, and clarifying advisory and advocacy roles.
4. There is a need of techniques for rapid responses to changing contexts of decision-making in water resources management.
5. There should be more reliance on sequential decision making, learning feedbacks and experimentation, instead of detailed policy schemes.
6. Efforts should be directed more at improving data base for decision making in water resources management and less towards preparing detailed long-term projections. The degree of detail of the latter should be compatible with the time horizon of a given plan.

It should be recognized that all these goals cannot be achieved without having appropriate institutions, responsive to environmental quality issues as well as to accelerating technological and social change.

BIBLIOGRAPHY

Biswas AK. 1981. 'Systems analysis applied to water management in developing countries: Problems and prospects', *Environmental Conservation*, vol. 8, no. 2, pp. 107–12.

———. 1988. 'Bhima revisited: Impact evaluation of a large irrigation project', *Water International*, vol. 13, no. 1, pp. 17–24.

———. 1988a.'Systems analysis for water management in developing countries: Constraints and opportunities', *Bulletin, International Commission on Irrigation and Drainage*, vol. 37, no. 1, pp.13–22.

———. 1988b. 'Sustainable water development for developing countries', *International Journal of Water Resources Development*, vol. 4, no. 4, pp. 232–42.

Biswas AK, Qu Geping. 1986. *Environmental impact assessment for developing countries*. Cassell Tycooly, London, 232pp.

Brown LC, Barnwell TO Jr. 1986. 'The enhanced stream water quality models QUAL2E and QUAL2E-UNICAS; Documentation and user manual', Environmental Research Laboratory, Office of Research and Development, US Environmental Protection Agency, Athens, Georgia.

Chankong V, Haimes YY. 1983. *Multiobjective decision making: Theory and methodology*. Elsevier North-Holland Publishing Co., New York.

Cohon JL. 1978. *Multiobjective programming and planning*. Academic Press, New York.

Davis RK. 1968. *The range of choice in water management*. The Johns Hopkins Press, Baltimore.

Eisel LM, Seinwill GD, Wheeler PM Jr. 1982. 'Improved principles, standards, and procedures for evaluating federal water projects', *Water Resources Research*, vol. 18, no. 2, pp. 203–10.

Frenkiel FN, Goodall DW. 1982. *Simulation modelling of environmental problems*, John Wiley, New York.

Fox IK. 1976. 'Institutions for water management in a changing world', *Natural Resources Journal*, vol. 16, no. 4, pp. 743–58.

Haimes YY, Hall WA, Freedman HT. 1975. *Multiobjective optimization in water resources systems: The surrogate worth trade-off method*, Elsevier Scientific Publishing Co., Amsterdam.

Haimes YY, Kindler J, Plate EJ. 1987. 'The process of water resources project planning: A systems approach', *Studies and Reports in Hydrology*, no. 44, Unesco, Paris.

Hamilton FEI. 1978. 'Some views on IRD' in *Strategy of future regional growth* (ed. M Albegov), Laxemburg, pp. 93–102.

Kaynor ER. 1978. 'Uncertainty in water resource planning in the Connecticut River Basin', *Water Resoures Bulletin*, vol. 14, no. 6, pp. 1304–13

Leopold LB, Clark FE, Hanshaw BB, Balsley JR. 1971. 'A procedure for evaluating environmental impact', Geological Survey Circular 645, U.S. Government Printing Office, Washington DC.

Liebman JC. 1976. 'Some simple-minded observations on the role of optimization in public systems decision-making', *Interfaces*, vol. 6, no. 4.

Major DC. 1977. 'Multiobjective water resource planning', Water Resources Monograph 4, American Geophysical Union, Washington, DC.

Miser HJ, Quade ES. 1985. *Handbook of systems analysis. Vol. 1: Overview of uses, procedures, applications, and practice*, North Holland Publishing Co., New York, 346pp.

————. 1988. *Handbook of systems analysis. Vol. II: Craft issues and procedural choices*. North Holland Publishing Co., New York, 681pp.

NZCAA, 1987. *Catchment authorities—Water and land resource managers of N.Z.* New Zealand Catchment Authorities Association.

O'Riordan J. 1985. 'Cumulative assessment and freshwater environment' in *Cumulative environmental effects: A binational perspective*, Canadian Environmental Assessment Research Council (CEARC), Ottawa, Canada, pp. 55–66.

Patten BC. 1975. *Systems analysis and simulation in ecology. Vol. III*, Academic Press, New York.

Quade ES. 1980. 'Pitfalls in formulation and modelling' in *Pitfalls of analysis* (eds. G Majone, ES Quade), John Wiley, New York.

USAID. 1980. *Environmental design considerations for rural development projects*. U.S. Agency for International Development, Washington DC.

White AU, White GF. 1978. 'Behavioral factors in selection of technologies', Proceedings of the ASCE (American Society of Civil Engineers) Convention, 16–20 October 1978, Chicago, pp. 26–46.

CHAPTER 4

General Management Techniques

$\diamond$

PART I

Techniques and Methods: Institutional, legal and financial framework

WANCHAI GHOOPRASERT

INTRODUCTION

In this chapter two areas will be considered: first, the techniques and methods that can be used by managers to ensure public involvement, and to manage human and financial resources; and secondly, the general institutional, legal and financial frameworks with examples of different types of institutional structures, project planning at various stages, budgeting and fund-raising.

TECHNIQUES AND METHODS

Public Involvement

One of the first tasks when launching a water resources project is to create awareness among the people in the community. This is to narrow the gap between the executing agency and the project area people. Also, the concerned persons will have the opportunity to reflect their own needs and to contribute to the means to fulfil the needs.

Public awareness and involvement could take a wide variety of forms such as:

- public opinion polls and other surveys
- the ballot box
- public meetings

- public hearings or inquiries
- representation of pressure groups
- task forces
- village volunteers, like in the case of rural water supply and sanitation projects.

Even though various media can be used, it was found that any form of newspaper or leaflet publicity is not effective, since literacy rate is usually low in communities in developing countries (Ghooprasert 1986).

Let us take the example of village volunteers. The volunteers should consist of both males and females, so that they can get along well with everyone in the community and are able to obtain frank opinions. Typical tasks of the volunteers include:

1. meeting with the people and local organizations such as the village headmen, temple abbots, youth groups, women's club, etc. It is important that they assess the real needs of the individuals that are free of any influence from others. Every sector of the community should be visited, since they may have different problems and opinions. It is, therefore, important to predetermine the best way to approach the villagers, e.g., door-to-door interviews, selected interviews, etc.
2. gathering the basic information of the community. Relevant information include the structure and organization, water resources, beliefs, economy, education, effective means of communication, technology and local expertise.
3. studying local behaviours.
4. conducting a briefing with the villagers to let them know of the findings, the problems and the needs. They should be allowed to voice their opinions and to participate in finding the solutions. A consensus should be obtained in this meeting.

If it is found through the meeting that the project being launched does not match with the needs of the community, agencies that are relevant to the needs should be contacted.

Participation at the village level usually depends upon the social structure of the community, the class composition, social stratification and status, hierarchy of the groups and the distribution of power and authority. A common pattern of participation is shown in Fig. 1 (Setty 1985).

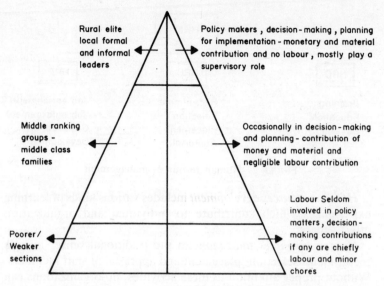

FIGURE 1. Common pattern of participation in a village

Human Resources Management

All organizations are faced with the necessity of using certain kinds of resources, such as physical, financial and human resources to meet their goals and objectives.

It is the people that make an organization a success or failure. Collectively they make an organization function the way it does. For an organization to function efficiently, its members should be properly trained and they should have appropriate experience. Training and having the right experiences take time. These cannot be inculcated into staff members overnight. Accordingly, good human resources management practices are important prerequisites to ensure that any organization functions at a high level of efficiency.

Human resources management (HRM) consists of three major areas: Human Resource Development (HRD), Human Resource Utilization (HRU) and Human Resource Environment (HRE). This is illustrated schematically in Fig. 2 (Nadler 1980).

Since it is difficult, if not impossible, for all organizations to obtain staff members who are all properly trained and have the right experience, appropriate steps need to be taken to ensure that employees develop the necessary expertise to carry out their tasks.

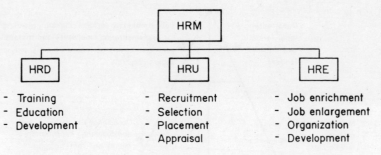

FIGURE 2. Human resources management

Human resources development includes various kinds of learning experiences which contribute to individual and organization effectiveness.

Human resources utilization, in the traditional sense, include recruitment, selection, placement and appraisal of staff members. Without proper attention to these activities, no organizations can continue to function effectively.

The third area, *human resources environment*, emphasizes the quality of the work environment.

Human resources planning traditionally has been within the preserve of human resources unit. They are the ones who should be in direct contact with the upper levels of the management in order to be fully aware of the long range requirements and aspirations of the organization. They must ensure that appropriate actions are taken well in time so that right quality of staff are available for the future operations of the organization.

The division dealing with human resources environment should also be involved since the kinds of people that are to be recruited and selected will have a direct impact on the identification of the appropriate environmental factors. Human resources development group has a contribution to make by providing the learning experiences to the necessary staff members so that a desired workforce is developed in a timely fashion. This is generally carried out by appropriate manpower planning and development techniques. Figure 3 illustrates the human resource planning process that is generally used.

For proper human resources planning for the water sector, initially a sectoral analysis should be carried out which could provide infor-

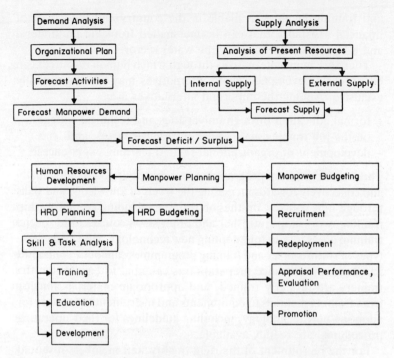

FIGURE 3. Process of human resources planning

mation on manpower demands for the short term (around 2 years), medium term (approximately 5 years) and long term (say 10–15 years). This analysis should identify different types of professional skills required and also their respective numbers. In addition to this investigation, a skill and task analysis of the available staff members must be carried out. On the basis of these analyses, strategies for human resources management can be developed. The units for which such planning studies can be carried out could range from an individual organization to a particular region to a country as a whole or even for a selected group of countries. Then policy decisions must be made as to how best to achieve these manpower requirements.

It should be noted that manpower development policies also depend on a number of external factors which could directly influence the water sector. Among these are the rate of growth of economic development, national planning priorities and policies, educational and training facilities available in the country and the extent of financial assistance that can be marshalled from the multilateral and bilateral aid agencies for the water sector.

There are many alternatives through which human resources can be properly developed. These alternatives may not be mutually exclusive. Among the important alternatives are:

- formal education through universities and technical colleges;
- on-the-job training and regular refresher courses; and
- development of organizational-related learning experiences.

These three different programmes will contribute both individually and collectively towards increasing the levels of knowledge and skills through new courses, on-the-job training including apprenticeship training, workshops refresher and other short courses, and special training programmes for adopting new technology. However, these types of study courses and training programmes cannot be conducted in a vacuum. Advance preparation is essential to ensure that the trainers are properly trained, and appropriate materials ranging from training manuals to equipments and instruments necessary for adopting new technology, including guidelines for their operating procedures, are readily available.

For the recruitment of the right quality staff members it should be ensured that the selection procedures are based on fairness to the candidates and the public at large, as well as the institution concerned. However, sometimes it is evident that too stringent an interpretation of the principles, when combined with bureaucratic red tape, could contribute to lengthy, cumbersome and costly recruitment arrangements. Thus, recruitment processes may need to be reviewed so that they could be further simplified and streamlined but at the same time ensure that right quality staff members are being recruited.

Manpower development policies must offer real opportunities for professional career development by objectively reviewing individual performances, providing promotional opportunities to all on a fair basis and systematic job rotation.

Promotional opportunities available to the staff members, in addition to an adequate and fair level of salary, are the two most important factors to attract and retain good staff members in the organization. Systematic job rotation is another important means available to increase the knowledge and capabilities of the younger staff members. Equally, this also can be used as a disciplinary measure. However, job rotation could entail some danger if a specialized individual is replaced by an inexperienced one for a demanding task. The reverse, where a specialist is appointed to a position where such skills are not necessary, could pose problems in terms of deteriorating staff morale.

Special incentives may be necessary in some cases where the tasks are difficult and/or strenuous, or if the assignments are in a remote or inhospitable region.

Delegation of authority is another important consideration for improving the quality and speed of decision-making at all levels. In many developing countries, the absence of proper delegation of authority is now leading to serious bottlenecks in a large number of water-related institutions. Because organizational management is done by human beings, performance appreciation, presence of a set of proper incentives and adequate career enhancement opportunity are important ingredients to ensure proper management. Finally, even with the best intentions, human resources management is unlikely to succeed if the institution does not have a clear staff policy which is not only considered to be fair but also is seen by the staff members to be objectively implemented.

FINANCIAL MANAGEMENT

Proper human resources management and financial management are absolutely critical requirements to ensure efficient functioning of any organization.

A budget can be considered to be a simply financial plan. Effective budgeting and cost control are crucial elements of financial management. There are several reasons to develop proper budgets, among which are:

- to encourage thinking ahead;
- to encourage coordinated thinking and communication;
- to develop standards for future performance;
- to integrate expectations about external factors and internal management policies;
- to identify problems before they occur and attempt to resolve them; and
- to anticipate change and adapt to it.

The budgeting process, as shown in Fig. 4, is a method to improve

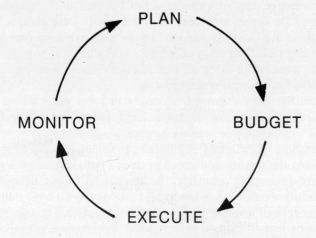

FIGURE 4. The budgeting process

operations. As the diagram shows, budgeting is a continuous process and requires a set of standards or targets. Budgets should be reviewed periodically and updated, if necessary, to ensure that plans and targets are being met. The budget should be regularly monitored so that the performance of the organization can be reviewed and evaluated.

Historically, budgets were used as a means to control expenditures. However, budgeting is currently often viewed as a tool for determining the most productive use of resources. Three types of budget are generally used.

(i) Appropriation budgets. These are often used by various government agencies or enterprises. It is based on the concept of spending up to a predetermined set of limits.

(ii) Flexible budgets. These are often used by private enterprises. It is based on the volume of the business conducted.

(iii) Cash flow budgets. It is used to predict the flow of cash into and out of the enterprise.

In order to ensure that budgeting is an effective tool, it is important to take into account the following potential problems:

– Budgeting goals may supersede the enterprise's own goals. It should be clearly understood that budgeting is a means to an end but not an end by itself.

- Budgets are likely to fail unless they are updated and readjusted regularly or as necessary.
- Budgets should not be based solely on historical data. This could repeat past inefficiencies, and may not respond to new demands and opportunities.
- Budgets defeat their purpose when they are used primarily as pressure devices. Budgets should be prepared in consultation with subordinates.

Budgeting begins with the statement of objectives of an institution. This would dictate the long-range budget requirements. A part of the plan may include an estimation of long-range performance expectations. Short-term budgets should be formulated within the framework of the long-term plan. They should reflect management expenditure financial policies.

Operating budgets are based on the projections prepared during financial planning. They should be prepared so as to provide incentives for improved efficiency. Well-prepared operating budgets provide management with the tool to control costs, evaluate actual performance, and identify and correct problems quickly.

Effective budget preparation requires the participation and commitment of staff at various levels. Every activity across the water sector should be represented including planning, design, construction, administration, operation and maintenance. It is not a good idea to allow budgets to increase on an across-the-board basis in terms of certain historical levels. Instead, each water resources manager should review the tasks assigned to him in terms of manpower, work load and resources, and then prepare the budget accordingly. Furthermore, each area of the water organization should assume direct responsibility for managing that part of the budget allocated to them, so that both the level of spending and the performance can be monitored on a regular basis.

After the budgets are finalized cost control measures should be organized. These measures are taken to compare actual financial performance against budgeted levels with respect to the timely achievement of the objectives. Specifically, cost control measures comprise regular reviews to identify variances and to stimulate required improvements. Thus, if there is a major cost overrun, it is necessary to review why and where the variances occurred. All the managers who were responsible should be involved in this review.

After the sources of variances are identified, it is important to determine why they occurred. Usually total variances result from more work completed than originally anticipated, while process for materials and labour remained the same. Price variances occur when prices paid were higher than budgeted, while the amount of labour and materials remained the same.

Once the sources of cost overruns are identified, managers should take necessary steps to correct them. Over time, as water resource managers gain more experience with budgeting and cost control processes, variances should decrease. The process of budgeting and cost control can then be regarded as an important constituent of financial management.

INSTITUTIONAL, LEGAL AND FINANCIAL FRAMEWORK

Institutional Structures of Water and Environment Management

It is impossible to generalize the organizational structure of water authorities that would be suitable for all developing countries. The structure will depend on many factors among which are the exact nature of work, size, total coverage, legal requirements, general environment, etc. However, works can typically be grouped into four or five categories as shown in Fig. 5.

Under the General Manager, a typical organization would divide the responsibility into say three groups. The Technical Services Unit is responsible for planning, engineering design and construction. The Operations Unit is usually responsible for the operations and maintenance of the water facilities. The Finance Unit is responsible for management of financial resources, accounting and budgeting. The Administration Unit takes care of fixed assets inventories and management, logistics, legal assistance, procurements and personnel

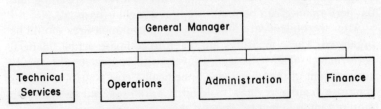

FIGURE 5. Typical organization structure of water authorities

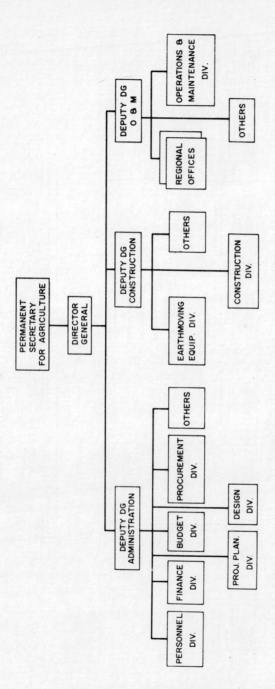

FIGURE 6. Organization chart of Royal Irrigation Department, Thailand

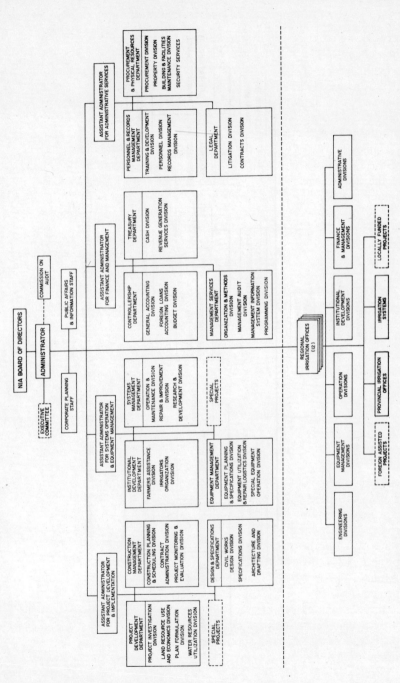

FIGURE 7. Organization chart of National Irrigation Administration of the Philippines

services including selection, recruitment, training and appraisal of the personnel.

Figure 6 is the simplified organization chart of the Royal Irrigation Department (RID) in Thailand. Here, under the Director General, there are three deputies; one for administration which includes the functions of finance, budgeting and engineering, one for construction and the third for operations and maintenance.

Unlike the case of Thailand, the National Irrigation Administration of the Philippines is divided into five areas under the leadership of the administrator (Fig. 7). The functions of engineering and non-engineering are clearly delineated. For instance, instead of having one assistant responsible for mixed areas such as the case of RID in Thailand, the tasks of planning, designing and contract administration come under the assistant for project development and administration, whereas financial and management matters come under the assistant for finance and management.

Both types of structures have advantages and disadvantages. For example, it is very difficult to find a person who could handle both engineering and finance. However, some types of work, even though of different nature, are closely related such as planning, budgeting and management information system. Separating and placing them under different lines could mean less coordination and longer flow processes.

In Africa, the work is usually assigned to different ministries as well. For instance, in Zimbabwe the Ministry of Local Government, Rural and Urban Development coordinates all water supply and sanitation schemes through the National Action Committee for the International Drinking Water and Sanitation Decade, which comprises representatives from three other ministries related to water. The Ministry of Energy and Water Resources Development is a service ministry which supplies water on a commercial basis to all government institutions. This ministry also operates village water supplies. The Ministry of Health is responsible for well drilling projects, and the Ministry of Community Development and Women's Affairs is responsible for community water supply and sanitation projects.

In Tanzania, the Ministry of Water is responsible for water supply programmes and also provides technical assistance to regional and district levels. It is also responsible for sewerage and drainage in urban areas, while the Ministry of Health is responsible in the rural areas. At the regional level, all development programmes are created by the Regional Development Directorates under the Prime Minister's Office. Programmes at the district level are overseen by the Ministry of Local Government and Cooperation.

BUDGETING AND FUND RAISING

Budget of water authorities usually come from three major sources: government subsidy, revenue generated from its own activities and external funding.

Extent of the government funding depends on the economic status of the country. However, for nearly all developing countries, budget availability from the governmental sources is limited. The macro planners have to decide the level of annual investment for each sector as well as their relative priorities in terms of distribution of funds. Some water projects may be accorded higher priority if they are directly in the productive sectors, or in the productive infrastructure sectors such as irrigation and hydropower, since developing countries always have to strive for the rapid growth of the economy.

Because of the multipurpose nature of the large-scale water development projects, revenue generated by the projects may not go to the project authorities but to the government's revenue collection process. For example, charges for electricity used by consumers, even though it is generated by the hydroelectric facilities of the project, seldom go to the water project management authorities. So far as revenue from irrigation water use is concerned, generally charges per unit of water is kept artificially low for socio-political reasons in nearly all developing countries. Even these low charges are not regularly collected for water projects. Thus, generally, the revenue generated by an individual project through its own activities is very limited and amounts to a small fraction of its total expenditure.

There are many external funding agencies (WHO 1985) that could help bridge the shortfalls in investments. Several developed countries offer bilateral assistance in the forms of grants and soft loans. However, most bilateral sources require that goods and services be procured from their own countries. Similarly, but on a much larger scale, multilateral lending agencies such as the World Bank and other regional development banks like the Asian Development Bank offer loans. Procurements, nevertheless, are open to all member countries. Both types of such financial assistance have their own advantages and disadvantages, and developing countries should decide on the most advantageous source of funding for specific water projects.

REFERENCES

Ghooprasert W. 1986. 'Social aspects in water supply and sanitation', Proceedings, World Water '86 Conference, London, United Kingdom.
Nadler L. 1980. *Corporate human resources development—A management tool.*
Setty ED. 1985. 'People's participation in rural development: A critical analysis', *The Indian Journal of Social Work*, vol. XLVI, no. 1.
World Health Organization. 1985. Catalogue of External Support Publication No. 7, Geneva, Switzerland.

CHAPTER 4 / PART II

Environmental Education, Training and Research

M. B. PESCOD

MANPOWER PLANNING

It is more difficult to identify true manpower needs in the field of water and environmental management than in many other sectors of development. This field is so broad as to defy precise definition and its manpower requirements so variable as to make their quantitative and qualitative assessment at any time almost worthless. The market demand for trained personnel in any country will depend on the government's attitude to environmental control, which will be influenced very much by the stage of economic development and the public's perception of environmental quality. In developing countries with an emerging awareness of the need for environmentally-sound development of natural resources and management of water systems, manpower planning in the sector should concentrate on the need to train a wide range of environmental professionals rather than attempting to estimate the numbers required in the future.

Both public agency and private company manpower needs should be considered in the planning process. Water and environmental authorities must be staffed with qualified any experienced manpower if they are to carry out their statutory functions effectively. Industrial companies must employ environmental specialists, as well as have environmentally-aware management, if they are to react responsibly to environmental control regulations. Increasingly in developing countries, there is a need for specialist environmental consultancies to provide the service to industry which will allow them to conform with ever more stringent environmental control legislation. As governments apply more pressure to improve environmental quality and implement environmental impact assessment procedures, it will become apparent that many of the staff employed in water and environmental management do not have the qualifications and training commensurate with their responsibilities. High standards of professionalism are required in the field of environmental assessment and control and each country must not only recognize this but must also be prepared to plan to create suitable education and training opportunities.

PUBLIC AWARENESS

In North America and most of Europe the public plays an active role in forcing investments in environmental improvement even though, in most cases, the public perception of an environmental risk bears little relationship its actual level. Environmental pressure groups are vociferous in opposing development schemes which they believe have adverse environmental impacts and are instrumental in forcing water and environmental managers to be more accountable for violations of environmental control criteria. Environmental quality monitoring data are increasingly made public and managers of water systems are therefore encouraged to make greater efforts to achieve environmental objectives at all times.

The situation in most developing countries is very different. Public awareness of environmental issues has not developed to the same extent because the main focus has been on economic development. In nearly all countries, poor people are more concerned with survival than with the preservation of environmental quality and tend to tolerate even highly-polluted conditions without much complaint. Generally, such poor people will only begin to start thinking about the quality of life when the basic necessities of life are met. Under these conditions, higher-income earners and the media have tended to accept more responsibility for environmental action. Even then, however, only the more obvious and serious cases are normally taken up. Few developing countries have formed environmental action groups that are non-governmental in nature. The general pressure to preserve environmental quality applied to planners and decision-makers in most industrially advanced countries has not reached their counterparts in the majority of developing countries. Even so, as might be expected, the situation is not homogeneous throughout all developing countries and in some, such as India and most South East Asian countries, there is now considerable interest in environmental issues and increasing investment in environmental control systems.

Public awareness of environmental matters has developed in advanced countries primarily as a result of the high profile given to such issues in the media. Television is a powerful tool and governments in developing countries could make better use of it in the attempt to provide the general population with greater access to environmental education. Illiteracy, which is a problem among the poor of many countries, could be overcome by adopting innovative approaches to television coverage of the subject in those regions where a high proportion of the population see television on a regular basis. No doubt, as politicians come to realize the importance of environmental concern as a political issue, all forms of media will be encouraged to devote more time to environmental reporting. Ultimately, environmental education must be introduced into school curricula so as to develop environmental awareness at an early age. Only through a better understanding of ecological issues by the general population will a balance be maintained between industrial development and environmental quality in developing countries. Greater public participation in decisions on development and the environment will help maintain a realistic balance and require water and environmental planners and managers to be more accountable for environmental quality.

Nevertheless even in some countries where the environmental awareness of the vast majority of the public is low, consideration of the environmental impacts of major development projects is increasingly a normal part of the planning process. The past efforts of multilateral and bilateral aid agencies and the requirements of lending agencies have led to environmental impact assessment being a regular procedure in the development of water projects. This is the case at present in countries like India, Thailand and the Philippines. However, in these and many other developing countries, once water systems go into operation, little attention is given to assessing the real environmental effects and managers are not often pressured by the public or government officials into being environmentally accountable.

ENVIRONMENTAL EDUCATION OF SPECIFIC PROFESSIONAL AND SOCIAL GROUPS

Many professional and social groups have an influence on the environment by reason of their job. Engineers, architects, planners and industrial managers are examples of professionals likely to interact with the environment and workers and politicians (both local and central government) are likely to be influential in environmental issues.

Taking the professional groups first, it is essential that their general education exposes them to environmental concerns. This will often be achieved by giving them access to environmental courses in their first degree. While this has happened in some cases in North America and Europe it is not yet common in developing countries. In places with a highly developed public awareness of environmental matters, the interest of undergraduates in obtaining a better understanding of environmental problems and solutions is already aroused by the time they enter universities. This interest can be satisfied by offering general courses which can be taken by students majoring in any discipline. In Europe, such opportunities are more limited than in North America, where the credit system allows greater flexibility to choose courses outside the major subject of study. Many universities in developing countries have academic staff who are knowledgeable about environmental matters and experienced in the study of environmental problems but, too frequently, they have not been invited to prepare general courses being offered to all students.

A more accepted approach in Europe is to include exposure to environmental studies as an integral part of specific undergraduate degree courses, although it must be admitted that this is less successful in Arts subjects. Science and Engineering degree courses more readily accommodate environmental subjects, particularly where the discipline clearly has a role to play in the environmental field, such as biology, geology and geography, or where the practice of the discipline will inevitably have an effect on the environment, such as civil engineering, urban and regional planning and agriculture. Very often, staff in such departments are involved in environmental monitoring, assessment and control and can easily produce specific courses and/or integrate environmental discussions into their normal course material. In this way, the relevance of their knowledge to environmental issues or the impact of their activities on the environment can be demonstrated. The need for curriculum development to accommodate such environmental exposure is long overdue in many developing countries even though qualified staff are available to achieve such changes.

Politicians may or may not benefit from this exposure during first degree courses, depending on whether they attend university. Workers in developing countries would not normally enrol in university courses and might be expected, like many politicians, to be totally insensitive to environmental problems. The only feasible way the ill-informed politician or the worker could improve his knowledge of environmental matters would be if he or she had access to short courses specifically designed for interested lay people. These could easily be offered by the science-oriented universities in most developing countries if there was a demand for them. It is not likely in the short-term that such short courses will be developed in this field when the whole business of continuing education is at a low level of activity in most countries. There is the possibility of individuals taking advantage of 'distance learning' environmental courses, such as those at the Open University in the UK (Porteous 1985), but this would require a high motivation for academic learning on the part of a worker and an admission of deficiency on the part of the politician, not qualities for which either group are noted.

Vocational training of technicians and skilled workers in the area of water resources is always necessary. Suitable training programmes must be designed on a national, or even sub-regional basis, they must be tailored to local requirements and customs and should be in the local language. This area of training, and also the technical education of potential workers in the water industry, has been difficult in most developing countries and remains a problem area. However, the remainder of this chapter will concentrate on the education and training of professional staff in water resources.

EDUCATIONAL ROUTES FOR ENGINEERS AND SCIENTISTS

First Degree Courses

As has already been mentioned, many graduates with degrees in basic scientific disciplines, such as physics, chemistry, botany, zoology, microbiology and geology, enter water and environmental management with little undergraduate orientation to this field of specialization. Nevertheless, many broadly-based environmental sciences or studies courses exist around the world. Concern has been expressed by employers about whether such a broad yet shallow coverage of this diverse subject is a suitable foundation for staff to be employed on scientific and technical work in water and environmental management. A survey of existing courses in environmental studies at universites and polytechnics in the UK, however, has indicated that, on the whole, most courses are grounded in basic disciplines with optional additions on aspects of environmental control. In spite of this, it is a market fact that environmental studies or sciences graduates are not very successful in acquiring employment in water and environmental management positions in competition with traditional science graduates.

Engineering disciplines with a role to play in water and environmental management include chemical, electrical, and mechanical engineering but the dominant technological discipline in the field is civil engineering. Many undergraduate civil engineering degree courses include a component of what used to be called 'municipal engineering', and which now is more commonly termed 'public health or environmental engineering', but many do not. When available, such subjects may be taken as a minor or major specialization within the degree course and help to prepare the graduate for entry into the field with a considerable background of relevant information of immediate value to an employer in the water sector. Study of the wider dimensions of environmental management are less common in civil engineering first degree courses. The lack of access to textbooks in developing countries will exacerbate any deficiency of instruction in environmental subjects.

Some more specialist courses in sanitary or environmental engineering do exist at undergraduate level and many of the European opportunities are included in a World Health Organization (1985) report on the training of sanitary engineers. The Civil and Environmental Honours Degree Course at the University of Newcastle upon Tyne is an example of a modified civil engineering degree course with a strong environmental bias. This course, accredited by the Institution of Civil Engineers, successfully blends the essential subjects in civil engineering (structures, hydraulics, CE materials, geotechnics, surveving and mathematcs) with the range of environmental subjects indicated in Table 1. The degree prepares its graduates for careers in the water industry, in air pollution control, in water management or in environmental consulting, manufacturing or contracting. Although similar courses can be found in North America and Europe, developing countries have been slow to provide such environmental educational opportunities.

One other approach which could be beneficial for educating environmental scientists and engineers in developing countries is distance learning. The UK's Open University provides opportunities for part-time degree qualification through distance-learning study

TABLE 1. Environmental Subject in the Civil and Environmental Engineering Honours Degree at the University of Newcastle upon Tyne

Year 1 *B Eng and* *M Eng*	**Environmental science and resources:** Physical and organic chemistry. Biochemistry, biology and micro-biology. Assessment of air, water and land pollution. Noise and thermal pollution. Assessment amd management of resources. **Introduction to ecology:** Eco-system types, components, biogeochemical cycles, population ecology, habitat, niche, diversity, trophic structure, stability, dominance. Terrestrial and aquatic ecology.
Year 2 *B Eng and* *M Eng*	**Environmental engineering I:** Water supply. Wastewater treatment. Solid waste management. Air and water pollution. Terrestrial pollution. Dispersion of pollutants. Case studies of toxic, organic and thermal pollution. Reclamation and reuse of land.
Year 3 *B Eng and* *M Eng*	**Environmental engineering II:** Treatment processes, engineering aspects of wastewater treatment, treatment and disposal of industrial wastes. Physical and chemical processes in potable water treatment. Design hydraulics and pipeline engineering. **Environmental engineering III:** Engineering aspects of unit processes in air pollution control. Noise abatement. Solid waste treatment and disposal. **Environmental modelling I:** Mathematical modelling concepts in the analysis of environmental systems. Statistical techniques and applications. Physical dispersion and degradation of atmospheric aquatic pollutants. Noise attenuation. Modelling of waste treatment processes.
Year 4 *M Eng*	**Environmental modelling II:** Interaction of environmental models and management. Organization techniques applied to pollution control, water supply, wastewater treatment, solid and hazardous waste disposal. Case studies. **Environmental management:** Environmental impact assessment. Management policies, strategies, legislation and economic aspects of control. Economic and social implications of growth, role of technology, alternative/appropriate technology. Institutional alternatives. Instrumentation and monitoring of techniques and the design of control systems.

over six years. Environmental management course profiles have been presented by Porteous (1985), and available evidence suggests that this is a convenient way for technician engineers and scientists to transfer to the professionally qualified stream. However, top-up environmental modules could be taken by scientists and engineers with first degree qualifications who wish to prepare themselves for employment in the water sector. Attendance at short courses is not considered to be a satisfactory form of education for professionals without an environmental degree background but can be useful for post-experience training and updating.

Postgraduate Courses

Many courses relevant to water and environmental engineering and management exist at the postgraduate level around the world. The array of courses available to graduates in developing countries is awesome and it is not surprising that a candidate will have great difficulty in choosing where to go. Some part of the responsibility for choice will be taken away from the individual by the funding agency or by the employer. The majority of candidates entering postgraduate courses outside their own country are supported by their employers or by bilateral or international funding agencies. Because of this, most sponsored overseas students entering environmental postgraduate courses' in Europe and North America are Government employees. Rarely do private companies finance their staff to proceed for postgraduate study, which means that the private sector in developing countries is not well prepared to respond to increasing pressure from Governments to improve environmental quality.

In a short chapter such as this it is impossible to do justice to all the postgraduate courses available around the world. The author might be excused for concentrating on UK courses, and particularly that at Newcastle, but wishes to draw attention to the availability of postgraduate Environmental Engineering courses at the Asian Institute of Technology in Bangkok. This Institution, where the author spent more than 12 years, has a high reputation in this field and offers full scholarships to able Asian candidates. Many other courses are available in developing countries, usually poorly resourced, but some of the Middle East and Gulf countries now provide well-resourced courses of high calibre. More quality courses should be developed in other developing regions of the world to allow more environmental specialists to be produced locally or regionally. Postgraduate MSc and Diploma courses relevant to water and environmental management are now offered in many institutions in Europe and the World Health Organization (1985) report reviews some of these, although the coverage is not fully comprehensive for all countries. The MSc course in Environmental Engineering at the University of Newcastle upon Tyne, but one example, has undergone almost continuous change since its inception and the outline of the current version is as follows:

Major subjects (required)	Minor subjects (two optional)
Water quality	Environmental engineering for developing countries
Wastewater engineering	Environmental control engineering
Water supply engineering	Solid and hazardous wastes management
Coursework	Air pollution control
Dissertation	Water quality modelling
	Environmental impact assessment

This menu of subjects not only allows a degree of specialization within the broad field of environmental engineering, but also caters for the large number of overseas students and the increasing number of home students interested in developing country technology. The course can be taken full-time, in 12 months, or part-time, in six-week modules, over a maximum of three years. Students failing to meet the entry standards required for the MSc might qualify for the postgraduate Diploma in Environmental Engineering over nine months.

IN-SERVICE TRAINING REQUIREMENTS FOR ENVIRONMENTAL PROFESSIONALS

Clearly, it will be to the advantage of employers in developing countries to send for postgraduate study those professionals who can return and train others 'in-house'. Only by taking advantage of the 'snowball' effect resulting from training the trainers will agencies with limited resources develop the quality of educated and trained manpower necessary to achieve improved environmental standards. In addition, many developing countries would benefit greatly from the introduction of in-service training requirements after first qualification and before acceptance as a professional scientist or engineer. Because this system is well developed in the UK, the author will review the further training requirements and opportunities for water and environmental engineers, scientists and managers in that country as an example of what might be appropriate.

Engineers

Engineering and other graduates who expect to become registered as Chartered Engineers (CEng) in the UK will be governed by the Engineering Council's 'Standards and Routes to Registration' (SARTOR). Present policy requires two stages of training beyond Stage 1 registration, which is on graduation from an accredited academic course.

Stage 2 registration will be achieved after successfully completing an accredited training programme or scheme designed: to develop the trainee's potential; to ensure that trainees develop their skills to the maximum for the benefit of employers, themselves, and the community; and allow trainees to achieve nationally recognized standards of competence. Throughout the training period, the trainee must maintain a record of work undertaken (which will be reviewed) and progress must be regularly assessed. In the case of the Institution of Civil Engineers (ICE), recognized as an Authorized Body by the Engineering Council, this stage of training has an additional requirement for attendance at three weeks of further-education courses and culminates in a professional examination (PE1) to verify technical competence. As a result of its Chilver Committee recommendations, the ICE scheme of graduate training is carried out under recognised supervising civil engineers, with the assistance of and monitoring by regional training officers.

Stage 3 registration, the final step towards CEng status, follows a period of responsible experience during which a candidate is required to maintain a certified record of work undertaken and responsibility carried and to submit to a professional review at the end of the period. The ICE also requires three weeks of further-education during this final period of qualification, which ends with the passing of a second professional examination (PE2) designed to test professional competence. Many engineers in developing countries are forced to take the professional examinations of advanced country institutions because professional bodies in their own countries have not taken on this training assessment role.

Scientists

A first or second class Honours science graduate wishing to become a Chartered Chemist or Biologist in the UK will be expected to have undergone a period of three years of satisfactory experience but a structured training programme is not specified. The changing face of technology forces scientists to look for further-education opportunities and employers are generally willing to support in-service training. Even in the UK, where the Water Industry Training Association (WITA) provides many training courses, there is a general feeling that not enough time and money are invested by the industry in continuing professional development. This situation is typical of that in most developing countries and agencies responsible for water resources development must become aware of the benefits of post-experience training for scientists and engineers.

Managers

Senior management development programmes, adopting a range of management training techniques, are operated in most large public water agencies in the UK, as well as by WITA. Increasingly included in such training programmes is leadership training, an important dimension of any manager in a declining manpower environment. Such training should also be beneficial for managers in developing countries where maximum benefit has to be gained from available resources.

RESEARCH

Appropriate Research

In recent years, much research on various aspects of water and environmental management in developing countries has been carried out around the world. Little of this has been properly structured or directed towards the major problems and most has been ad hoc investigation of topics chosen to suit the interests of individual researchers. In developing countries, too much research effort has been directed towards such esoteric themes as the 'kinetics of the activated-sludge process' rather than to the crying needs of local problems with much more bearing on the quality of life of large numbers of people. Graduates returning from developed countries with research experience have been imbued with the idea that only fundamental scientific research (in a very narrow sense) is academically legitimate and it is not surprising that they follow the direction of their supervisor's research interests, at least until they themselves gain some international reputation. In the past, the lack of interest shown by organizers of conferences and publishers of technical journals in research applied to the relatively simple problems of developing countries has been another reason for them to pursue more marketable theoretical topics. Technology and information transfer between developing countries has generally suffered as a result of these deficiencies.

More recently, international conferences have regularly included papers on the problems of developing countries and it has become scientifically acceptable to concentrate on the more immediate applied research needs in these countries. Conferences have been devoted entirely to discussions of water and environmental issues in developing countries, and international agencies have become involved in, and are encouraging, research applied to major problems in this area. Although the number of skilled researchers in this field is increasing in developing countries, there is little awareness on the part of their governments of the need to sponsor water and environmental research. Where national budgets are strained to support the conflicting demands of different sectors in a developing economy it is essential that public funds are applied efficiently; only by financing research can water management and environmental control be achieved in an optimal way at any particular stage of development.

A Balanced Technological Approach

Some water and environmental problems in developing countries do not call for sophisticated solutions, but it should not be tacitly assumed that the answer for all situations lies in the application of what has become known as 'appropriate technology'. Neither should it be presumed that technical manpower to carry out sophisticated research and development is not available in developing countries; many well-trained and highly capable professionals can be mobilized in almost all developing countries, and their talents can be usefully applied in optimizing simple systems. Several approaches to water and environmental research appear to be apposite in the context of developing countries:

- the application and adoption of known principles to specific local problems;
- the development of low-cost, simple technological systems for general use; and
- the optimization of simple systems and their introduction using

sophisticated techniques (e.g. mathematical modelling and computing) whenever appropriate.

There will always be situations in developing countries where a sophisticated solution is indicated and affordable, such as perhaps the provision of sewerage in the central business district of a major city, but research will not usually be called for. In such cases, it is more economical to adopt standard approaches from countries where these have been documented and the necessary expertise can be bought in initially, as is often the current practice with the relatively few projects for which there is finance.

The more critical area for research is where conventional solutions cannot be afforded and where local conditions require innovation. However, in designing simple systems, the opportunity for staged upgrading over time must not be overlooked. It is as irresponsible to limit the opportunity of people in developing countries to improve their environment along with development as it is to think in terms of providing the few with advanced technology immediately at the expense of the many without even basic systems. Large urban centres and large-scale water projects will require the full range of technology in water management and environmental control and choosing the right balance for the time, while designing for integration of all components into a long-term masterplan, is essential from the social and economic points of view.

Social, Economic and Institutional Perspectives

Although there are research needs of a purely technical nature in water and environmental management, all developments must stand up to social and economic evaluation. The social, cultural and religious preferences of the beneficiaries must be taken account of if a technical solution is to be implemented, and it is very important to incorporate social-survey findings into decision-making at an early stage of research. Economic constraints will always be a factor in water and environmental research but, in relation to developing countries, special consideration must be given to the aspirations for rapid economic development and to the conflicting demands on public funds from industrialization and from environmental control.

Increasingly, a community's 'willingness to pay' and 'ability to pay' for improvements are being adopted as economic criteria in water and environmental programmes associated with human settlements, acknowledging the inability of government revenues to cover the full costs of such programmes. The importance of involving the communities which will benefit from technological research at the beginning of projects becomes even more obvious when they are not only the recipients of the results but will also be expected to pay for them—genuine consumers in a market sense. A multidisciplinary team of researchers will be in a better position to consider all factors than a team composed only of engineers and scientists.

Extending the same concern for acceptability of the results of research in the real world, another group which needs to be convinced of the suitability of research ideas is the many government officers in agencies likely to be involved in the implementation of water and environmental programmes. Unless key figures in the relevant agencies are kept fully informed of the research, and made to feel as if they are indeed part of it, no research team can expect their recommendations to go beyond the final-report stage. Full collaboration with the responsible agency will help to ensure that the laboratory, pilot-scale or desk study will be translated into at least a demonstration project in the field. Very often, local agency funds will be made available to support such research activities, and international organizations will find it easier to support this type of collaborative research and development work. Involvement of government staff in research projects will simplify the social survey, which should form an integral part of water and environmental management research.

Support for Research in Developing Countries

There can be no doubt that the best place for research on water and environmental problems in developing countries is in those countries themselves. A certain amount of research can be carried out in advanced countries, but this is limited by the difficulty of simulating the correct environmental conditions. Shortage of qualified staff can no longer be a reason for not undertaking on-site research, although the rewards available in developing countries tend to prevent the most competent researchers from returning to their homes. University salaries are often very low and otherwise able water and environmental researchers are forced to take on design consulting work and second jobs, thus limiting the time they can spend on research. Government establishments for environmental research do not exist in many developing countries, although sometimes a section of a central industrial research institute is concerned with this field. Again, governments are not usually generous in their payment of research workers and, with a few exceptions, little valuable research has been produced in these institutions.

Financial assistance for water and environmental research has always been a limiting factor in the past. However, the activities of international organizations over the past decade, culminating in the UN Drinking Water Supply and Sanitation Decade, 1981–90, have channelled funds into research on the infrastructure of human settlements in developing countries. Many UN and bilateral agencies have been, and continue to be, involved in these research and development programmes but they have concentrated primarily on water supply and sanitation and have thus benefited only the few researchers in this area.

Some programmes have been organized on a global or regional basis to include research projects in various countries, with the great advantage of the research results being widely circulated and probably more generally applicable. Examples of this type of coordinated research are the World Bank's programme on Appropriate Technology for Water Supply and Sanitation in Developing Countries, the WHO International Reference Centre for Community Water Supply's programme on Slow Sand Filtration and the Canadian IDRC's programme on Water Supply and Sanitation for Rural Areas and Squatter Settlements. These particular research programmes have provided respectability as well as finance to workers in this low-technology field. They have also focused attention on the importance of economic, environmental, health and socio-cultural effects of technologies in identifying the most appropriate systems to suit the needs, preferences and resources of difference areas. There is now a need to expand such support into the broader areas of water resources development.

The Importance of Demonstration Projects and Evaluations

Although a research project might identify a promising approach in the water and environmental management field it will not often be taken up by the government agencies concerned unless it is proved in a field demonstration. This will be particularly so for smaller scale systems, on-farm water management schemes, etc.

but will not generally apply to large scale water developments. Government officers are unwilling to risk their reputations on new systems, especially if they involve low technology, because failure is not looked upon kindly in any country, even if an attempt was being made to reduce the costs of providing a service. Where appropriate, demonstration projects are ideal for testing the validity of research findings in a real situation and yet do not commit the agency to large-scale investment. They also afford the opportunity to test the operational and maintenance characteristics of a system, and to train agency manpower. It may be necessary to convince a bilateral or multilateral funding agency that a new approach, developed from research, is technically and economically viable and the demonstration project has been widely adopted to justify large-scale investment.

Sometimes, a demonstration project related to human settlements will have a catalytic effect in spreading the demand for an improvement among neighbouring families or communities, and this is critical if communities are expected to contribute when the full-scale programme is implemented. Convincing the public that a development is beneficial is an essential element in water management and environment control, and demonstration projects can be adopted as a means of informing the wider community. Another possible use of demonstration projects is as workshops to encourage and direct community participation in the follow-up programme. Occasionally, it will be possible to develop the capacity of an 'informal' sector of the community to scheme by extending a basic demonstration project to include training and finance for such a purpose. The potential for doing this is dependent upon the willingness of authorities to relax building codes and quality standards, and so to accommodate possibly inferior but nevertheless serviceable items produced by the 'informal' sector.

A most important aspect of research in developing countries is *ex-post-facto* evaluation of completed large schemes and of demonstration projects. Too often in the past, designs have proved too costly or too sophisticated to operate and maintain, and have been abandoned. Examples of many 'white elephants' exist in the developing world, and remain as monuments to the misguided designers who, no doubt, never looked back on their masterpieces. Only by careful evaluation of completed projects or programmes can the agency, the designer and the researcher be made aware of good and bad aspects of the design and so prevent the perpetuation of mistakes. Sometimes the success of a particular project can be improved by *ex-post-facto* evaluation, if it identifies, for example, a deficiency in the social acceptance or understanding of a development and recommends further public education. The adequacy of administrative arrangements for operating and maintaining a system and collecting essential revenues will also be exposed by such evaluations.

For large-scale projects, the actual impact on the environment and on the community is often overlooked once the project goes into operation. Even when an EIA has been carried out at the project planning stage, follow-up studies are rarely carried out. This evaluation type research on completed water resource schemes will require careful planning and the involvement of multi-disciplinary teams. However, it will be highly cost-effective in preventing mistakes to be perpetuated in future projects and allowing adverse impacts to be counteracted in the future and overcome in the present.

REFERENCES

Porteous A. 1985. 'A degree profile in environmental management at the Open University', *Wastes Management*, vol. LXXV, no. 12, pp. 698.
Royal Commission on Environmental Pollution. 1985. *Eleventh Report: Managing Waste—The duty of care*, (Chairman, Sir Richard Southwood), Her Magesty's Stationery Office, London.
World Health Organization. 1985. *Training of sanitary engineers in Europe* (ed. RB Dean), Regional Office for Europe, Copenhagen.

CHAPTER 5

Monitoring and Evaluation of Irrigation Projects

ASIT K. BISWAS

INTRODUCTION

There has been considerable controversy in the recent past over the desirability and efficiency of irrigation projects. The proponents of irrigation projects have pointed out that implementation of such projects not only increases the total food production of an area when compared to rainfed agriculture but also significantly improves the reliability of the production process by ensuring proper water control. It has been estimated that even though only about 20 per cent of the world's agricultural land is irrigated at present, this accounts for 40 per cent of the global agricultural production. In addition, irrigation provides the basis for a better and more diversified choice of cropping patterns and growing of high value crops, which otherwise may not have been possible. The water control structures built for large-scale irrigation projects often also simultaneously generate hydroelectric power and control floods, which further add to the national economic development process. Since globally irrigation is the largest consumer of water (about 80 per cent of all water used), and hydropower generation does not consume any water, these two are compatible uses of water.

On the negative side, vociferous opponents have claimed intense disappointments with the results of irrigation projects during the past two decades, due to high costs of projects, cost and time overruns, poor management, non-realization of full planned benefits, adverse environmental and health impacts, and exacerbation of differences of the existing social and economic distribution of assets amongst the farmers.

An important reason as to why irrigation projects are currently generating simultaneously both extreme optimism and pessimism is because of the absence of regular monitoring and evaluation processes that can clearly and unambiguously identify the impacts of the projects.

MONITORING AND EVALUATION OF IRRIGATION PROJECTS

Monitoring and evaluation have received much lip-service in the present decade, but have seldom been carried out properly and effectively. Indeed besides some rhetoric, one would be hard-pressed to identify a single irrigated agriculture development project that has been monitored and evaluated properly and regularly, and where the results of monitoring and evaluation are used to improve the management of irrigation projects in order to ensure that the expected benefits do accrue within the planned time horizon. This state-of-the-art review is based on the author's experience as an advisor to 17 developed and developing countries and all major international organizations involved with irrigation.

In the present context, monitoring may be defined as continuous or periodic surveillance over the implementation of the necessary irrigated agricultural activities, including their various components, to ensure that work schedules, input deliveries, targetted outputs and other required actions are progressing according to the plan. Since the primary purpose of monitoring is to achieve efficient and effective project performance, it should be considered to be an integral part of the management information system, and thus should be a regular internal activity. Evaluation may be defined as a process which determines systematically and objectively to the extent possible, the impact, effectiveness and relevance of project activities in terms of their objectives.

There are two points worth making here. First, evaluation can broadly be divided into two broad categories: ongoing and periodic. Ongoing evaluation can be used to examine whether any changes are necessary for the operation and management of a project to ensure that its performance is satisfactory and the overall objectives can be achieved. For example, it is possible that in certain cases the assumptions underlying the irrigation design may have been inappropriate so that the farmers are not receiving their expected share of water. If through the monitoring and evaluation process this or similar types of problems can be identified, the question arises as to what measures can be taken to rectify such situations.

In contrast to ongoing evaluation which is a continuing activity, periodic evaluation is carried out after longer time intervals, say every five years or so. Periodic evaluation generally deals with achievement of socio-economic objectives, which may not show any discernible or significant change over a shorter period of time, say one year or so.

The second point is with reference to the base with which project impacts or changes can be compared. Very often reliable socio-economic data on pre-project conditions in developing countries do not exist. Furthermore, project objectives as originally outlined may be fuzzy, inaccurate or may not be sufficiently quantitative for evaluation purposes. Many times objectives may require re-definition or sharpening of focus in the light of experience gained since a project was initiated. Thus, for evaluation purposes, blind adherence to initially stipulated objectives may be counter-productive.

It can be persuasively argued that if the existing performance patterns of irrigated agriculture projects in achieving their objectives are to be improved significantly, it is essential to ensure that monitoring and evaluation become an integral component of the management process in order first to determine their achievement levels and then to identify what adjustments and corrective actions may be necessary to ensure that the future stream of benefits accrue in time to the appropriate target groups.

Some form of monitoring and evaluation is always done for irrigation projects. For example, most project authorities monitor the flow rates in the main irrigation canals but they may not have corresponding data on watercourses. Often losses occurring in watercourses are not known, and if such losses are higher than designed values, reliability of water availability to the farmers at the tail of watercourses declines significantly. In addition to certain irrigation factors like flow rates, some agricultural aspects such as crop yields may be monitored, but even these may not be monitored on a regular basis. When one moves from physical factors to socio-economic factors, the status of their monitoring and evaluation gets even worse. One would indeed be hardpressed to identify one single irrigated agriculture project where its impacts on the lifestyles of the intended beneficiaries have been evaluated at regular intervals.

Regular and reliable evaluation of irrigated agriculture projects, however, is not an easy task under the best of circumstances. There are methodological problems which need to be resolved in order to find a cost-effective and reliable approach that can be used for the evaluation of a specific project within the resources and expertise available. Even when methodological problems could be resolved, there are other important barriers like institutional inertia, and sometimes even downright opposition, which have to be effectively overcome before a serious evaluation can be undertaken which could be a part of the management process.

While limited literature exists on the integrated monitoring and evaluation of irrigated agriculture projects (Biswas 1987; Sagardoy 1985), unfortunately more reports are available on pseudo-evaluation or superficial evaluation that have been carried out in the recent past at both national and donor agency—bilateral and multilateral—levels, which are more concerned with the protection and enhancement of the reputation of the organizations concerned, both within and outside countries, and the individuals associated with the projects. These types of evaluations cannot have a beneficial impact on the management process since they either do not identify major problems and bottlenecks, or if they do, their importance is significantly downplayed. Such evaluations are not only detrimental to the society in the long run but also reduce the effectiveness of irrigation projects as well as the perceived usefulness of the monitoring and evaluation process.

A comprehensive and objective analysis of the experiences of the United States Agency for International Development of irrigation projects supported by them in various developing countries indicated that monitoring and evaluation practices of both donor and recipient countries have 'come in for criticism from each group about its own organization and about the activities of its counterpart', and that 'too little gets done by either group' (Steinberg 1983).

In view of unprecedented controversies in recent years on the efficiency and even the desirability of irrigation projects, it is essential that objective evaluation be considered mandatory, both to get reliable status reports on the operation of and benefits from the schemes and use the results to improve the management processes further in order that impacts on beneficiaries are maximized.

Why Monitoring and Evaluation?

There are many reasons to carry out systematic monitoring and evaluation of irrigation projects; the principal ones are the following:

– to determine the extent of achievements of the goals of a project

by assessing actual impacts and then comparing them with expected impacts;
- to obtain information as to why a project may not have had anticipated impacts by identifying the magnitude, extent and location of the problems in order that corrective actions may be taken to maximize the beneficial project impacts;
- to increase the understanding of the management of the various inter-linked processes and issues involved so that the resulting enhanced management understanding can be translated into action in terms of immediate, observable, concrete decisions;
- to verify the relevant project assumptions;
- to learn lessons to improve planning, implementation and manage- ment of similar projects elsewhere;
- to plan later phases of the project more effectively, based on the evaluation of the performance of the first phase;
- to contribute to the modification of the organizational behaviour on the basis of relative successes and failures of projects at various levels;
- to provide facts and success stories at the ministry or department level which can not only defend existing policies and programmes but also may assist in getting additional financial support; and
- to provide national policy-makers with objective information in order that they can decide to what extent such activities can be continued in other parts of the country.

It should be noted that the reasons outlined above for carrying out monitoring and evaluation are not mutually exclusive since they are often interrelated. Equally, it is not enough to identify and analyse the technical, social and economic aspects of the various issues and problems; it is essential to know about institutional arrangements and that constraints be reviewed as well, since it is the institutions concerned which in the final analysis have to develop ameliorative policies and implement them.

On the basis of reviews of irrigation projects in many countries, it is clear that no one is satisfied with the present status of irrigation planning and management in terms of monitoring and evaluation. The problems appear to be many-faceted, among which are the following.

Decision-makers claim that they have no clear idea as to what real impacts the irrigation projects have had on the anticipated beneficiaries, or even on the nature and extent of the real beneficiaries.

Planners point out that they have no objective information on how past planning of irrigation projects has fared, and without any reliable feedback they cannot improve the existing planning process.

Managers state that they cannot make appropriate timely decisions since the information they receive is generally unusable and of little help (too little, too much, irrelevant, unreliable or too late).

Engineers and administrators feel that they are already overworked due to routine administrative chores, preparing numerous reports which are administratively necessary but very few people appear to read, their required presence at numerous unproductive and questionable meetings, and other similar tasks which do not leave them with much time to carry out their real functions.

Evaluators feel that their works do not receive proper attention, and what is more they are not given enough resources or time to carry out their tasks efficiently.

Officers of funding agencies complain that they have no clear idea on the effectiveness of the projects and accordingly they have difficulty in making correct decisions on the funding of new projects.

A realistic assessment of the current status of monitoring and evaluation of irrigation projects would indicate that virtually everyone associated with such projects feels that evaluations are essential and beneficial, but in reality the rhetoric overwhelmingly exceeds actions. Organizations appear to embark more upon planning exercises and training programmes on evaluation than carry them out properly. This unsatisfactory situation must be improved.

MONITORING AND EVALUATION REQUIREMENTS

There are some fundamental requirements for designing any monitoring and evaluation system for an irrigated agriculture project. Among the primary requirements are the following:

1. timeliness;
2. cost-effectiveness;
3. maximum coverage;
4. minimum measurement error;
5. minimum sampling error;
6. absence of bias; and
7. identification of users of information.

Timeliness

Most management decisions have a time dimension, even though the timeliness of making some decisions may be more important than others. For example, if farmers at the tail ends of watercourses are not receiving their share of irrigation water regularly, or if fertilizers and pesticides are not available at the right time of the cropping season, it is necessary that immediate remedial measures are taken. If not, it would result in poor harvest, and the income foregone by the farmers will never be recovered. Thus, it is essential that information collected reaches the appropriate decision-makers on time so that rational decisions can be made in time based on the data monitored. Accordingly, for a rational management system, monitored information should be channelled in a timely fashion so that it can be converted into decision and action.

It should be noted that the management success depends not only on the timeliness of the information but also on the quality, extent and the form in which the information is channelled into the decision-making process. A problem often arises because even if the required information has been collected, it could not be channelled into the decision-making process since it is either in a diffused or inappropriate form or could not be obtained and analysed within the timeframe by which decisions should be made. In a review of monitoring and evaluation systems of irrigated agriculture projects (Biswas 1987), it was found that in many schemes pre-season crop-related information was not even analysed before the end of cropping seasons. Under such circumstances, monitoring and evaluation can only have a limited impact on the overall performance of projects.

The danger is that if monitoring and evaluation information from the project does not reach the managers on time, it is likely that one or more of the following consequences, which are not mutually exclusive, will occur:

- wrong action may be taken;
- decisions taken may not be optimal on a long-term basis;
- no action may be taken when one is desirable;
- decision taken may result in irreversible damage; or
- decision taken may unnecessarily increase the cost and timeframe required for the resolution of a specific problem.

It is therefore essential that a monitoring and evaluation system for an irrigated agriculture project be set up in such a fashion that relevant information in usable form reaches the people who need it on a continuing and timely basis.

Cost-effectiveness

Information collection, processing, analysis and scaling requires financial resources, expertise, manpower and equipment. Since the ready availability of all these resources in developing countries is limited, any monitoring and evaluation system designed for irrigated agriculture should be cost-effective. This essentially means a sensible trade-off between the depth and context of information to be collected as well as between amount, relevance and accuracy. As a general rule, it can be said that the overall value of information collected in terms of use should exceed the cost of obtaining that information.

For most projects, from a management viewpoint at any specific time, the value of information generally increases with the increasing extent and the accuracy of information available. Value of information, however, for most decisions generally approaches a plateau at a certain stage, beyond which it increases only marginally. In contrast, the cost of obtaining information continues to increase with more coverage and higher accuracy. This is shown diagrammatically in Fig. 1.

The shaded area in Fig. 1 is the cost-effective zone, beyond which the cost of obtaining information will rapidly exceed its intrinsic value. Exactly where within a shaded area a decision should

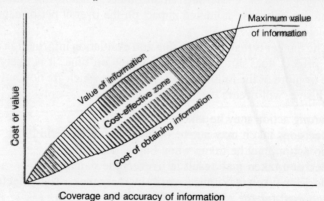

FIGURE 1. Cost-effectiveness of monitoring and evaluation of information

be made depends on a variety of factors such as type of projects, management experience and potential impact, but such trade-off considerations are often made on the basis of value judgements.

There is often a tendency to collect more data than necessary. For any monitoring and evaluation process to be efficient and cost-effective, it is essential to have a clear idea about who is going to use the data, what types of data are necessary, how would the information be used, and when and in what form they should be made available. Without such a clear focus, unnecessary and non-essential data may be collected, which is an expensive luxury all countries can do without.

While data collection processes need to be carefully integrated within the monitoring and evaluation framework, consideration and desire for scientific rigour should be balanced with the need for timely information. This in reality often translates into a trade-off between accuracy and reliability with cost. In the final analysis, the value of information for the decision-making process becomes the key factor to resolve this trade-off.

Maximum Coverage

For monitoring and evaluation of irrigation projects, especially large ones, a major difficulty arises from the fact that a wide area may need to be covered, wherein it may be necessary to get an objective view of the operation of the irrigation system and its impacts on agricultural production and the overall quality of life of the people in the area. Thus, maximum coverage taken literally may prove to be an expensive, complex and time-consuming process.

As a general rule, sociologists and anthropologists prefer to have as much breadth of coverage as possible. However, given high resource costs of manpower, time, transportation, and other related factors, as well as high opportunity costs, a decision may often be necessary to have limited coverage on selected variables, and then use the balance of available resources to obtain more detailed information on specific aspects and/or areas that are critical from a management viewpoint. Accordingly, maximum coverage in the present context should be interpreted to mean collection of maximum data that are necessary and will be used for management purposes, subject to resources (funds, manpower, expertise, equipment and time) availability.

Minimum Measurement Error

The level of accuracy and reliability of any data that will be collected is an important consideration for any monitoring and evaluation process. Generally engineers and physical scientists are more concerned with the accuracy of measurements and data collected than sociologists and anthropologists. For irrigated agriculture projects, measurement error could be a real problem when small farmers and landless labourers are being considered. They are often illiterate and may have some difficulty with precise numerical quantifications. Accordingly, they may not be reliable or even somewhat vague about the rate of changes, especially when the changes are within the order of 15 to 20 per cent and some time has elapsed between the two periods in time which are being monitored. The enumerators and data collectors should be aware of this potential problem and attempt to ensure that the changes are properly reflected.

Minimum Sampling Error

Since it is neither necessary nor desirable to monitor all possible developments in the project area, sample surveys are essential. Irrigated agriculture projects cover numerous issues and diverse disciplines, and accordingly there is no straightforward or uniform solution as to what may constitute suitable sample size. For example, for analyses of rainfall, one rain gauge per square kilometre will be considered to be a very dense network, and thus totally unnecessary unless very exceptional circumstances warrant it. In contrast, the identical sample size would be totally unacceptable to sociologists. In the final analysis, determination of sample size will depend upon the type of information to be collected, and the use that will be made of it.

Absence of Bias

Monitoring and evaluation of irrigated agriculture projects often suffer from biases of people performing the task. This happens because evaluators, often due to their disciplinary orientation, expertise and past experiences, may have the tendency to concentrate on specific issues at the cost of other issues which may be of similar importance. Among the common biases observed are water entering watercourses but not losses, irrigation but not drainage, fields near roads along canals and watercourses but not those to which access is difficult or uncomfortable, review during healthier, better-fed dry season when climate is pleasant but not during food-scarce, unhealthy, unpleasant and wet season, and interview large farmers or men but not small farmers or women. Equally, there is a danger that biases may be introduced in terms of one's discipline, since unidisciplinary people often tend to concentrate on areas that are of primary importance to them.

It should be noted that in a real world, an issue is an issue. It is often labelled engineering, economic, social or legal depending upon an individual's discipline, experience, and ways and means of approaching it. Thus, ideally, evaluations should be carried out by multi-disciplinary people, who may specialize in one discipline but are knowledgeable in other disciplines. They should be flexible, observant, sensitive, eclectic and constructive. They should be capable of intermixing freely and questioning sympathetically and inventively. Since, in reality, such qualified and experienced individuals are very difficult to find, one may have to depend on whoever is available. To a certain extent, the problem can be resolved by carefully choosing a multi-disciplinary team, which may offset biases of the individual members by the juxtaposition of the insights of various disciplines. Past experiences indicate that use of multidisciplinary teams for monitoring and evaluation of irrigated agriculture projects, where team members are not familiar or do not have established working relationships with each other, generally do not produce an integrated multi-disciplinary approach or report.

Identification of Users of Information

If the results of any monitoring and evaluation are to be actually used, it is necessary to identify who are going to be the users of information and their information requirements before designing a monitoring and evaluation system. At the different levels of management, the hierarchy of information needs is different. For example, at a certain level of management, detailed information on a specific aspect of an irrigated agriculture project may be necessary, whereas at other levels (generally higher), aggregated information may be required. It is necessary that the right type of information is provided to the appropriate levels.

For any utilization-focused evaluation, after identification of relevant information users, it is desirable (i) to actively involve the users in ways that would increase their commitment to the utilization of evaluation finds; (ii) to train users to increase their understanding of evaluation and make it possible for them to play a useful role in the evaluation process; and (iii) to provide genuinely useful information to the users so as to reinforce their future commitments to evaluation.

Trade-off between the Requirements

The principal requirements discussed above should not be considered individually in isolation since some may reinforce each other and thus are mutually supportive but others may be in conflict. The quality of any monitoring and evaluation system is determined by not any one of the requirements but rather how effectively all these factors are integrated in one system. For example, there is always a trade-off between maximum coverage, minimum sampling error, minimum measurement error and cost, and these trade-off decisions are generally case specific. There is no universal clear-cut solution.

There is sometimes a tendency to emphasize one or more of the requirements at the cost of others because of bias. A good example of this is the monitoring and evaluation of agricultural projects and programmes in Nigeria during the past 10 years that were funded by the World Bank (Lai and Felton 1986). Massive resources were devoted to monitoring and evaluation activities, and enormous data were collected by the Agricultural Projects Monitoring Evaluation and Planning Unit (APMEPU), which was established in Kaduna in 1975 to coordinate the monitoring and evaluation activities of the various Agricultural Development Projects/Programmes (ADP). In spite of such intensive efforts by APMEPU, the impact of the ADPs on food production or consumption is far from clear. A major problem arose because of APMEPU's overriding emphasis to minimize sampling errors. This contributed to a lop-sided approach which gave high priority to statistical considerations but low priority to other requirements and consideration of resources available to perform all the monitoring and evaluation tasks. While the sampling error was minimized, all types of other errors were introduced at a level that was unacceptable. This meant data messaging, further analysis and re-analysis, which not only took time but also contributed to the development of a credibility gap between the Unit and project management and other users of information. It is thus essential that a cost-effective monitoring and evaluation system be developed that provides information required by managers in a timely fashion, subject to resources and manpower constraints. Development of an effective system is an evolving process that requires regular, good feedback between the monitoring and evaluation unit and users of the information.

In addition to the requirements discussed, it should be noted that in a real world compromises have to be made, like in an ideal marriage, between the requirements identified, empirical techniques used, various professional expertise available, and practical constraints like cost, time and political and institutional factors. This may even require the consideration of a second-best monitoring and evaluation package that may be more attuned to what is possible in practice. However, in such a pragmatic approach, practical considerations should not justify the dismissal of technical requirements: on the contrary technical factors may become even more important when one may be forced to use a less desirable approach.

FRAMEWORK FOR MONITORING AND EVALUATION

Monitoring and evaluation of irrigated agriculture is a complex process since a large number of regular and specific tasks have to be performed, both concurrently and sequentially, in a coordinated manner, by a variety of professionals, within available time and resource constraints. Furthermore, potential decisions made by local, national, regional and international institutions may have a direct bearing on the project.

Three issues are worth noting here. The first pertains to the physical boundary within which monitoring and evaluation should be conducted. While hydrological or political boundaries may be comparatively easy to define, they are not the boundaries within which all project benefits and costs, both direct or indirect, are confined. Generally most of the project-related physical variables are confined within the project boundary, but aspects like migration, trade or environmental impacts often have implications far outside the project area. Accordingly, it is not an easy task to define the boundary within which impacts should be evaluated.

The second issue is the time dimension, which is another complicating factor. Some impacts are immediate, visible and quantifiable, and thus can be identified and considered within the monitoring and evaluation process. A few other impacts, however, may be slow to develop, and thus may not be easy or possible to monitor meaningfully in the early stages. For example, some unanticipated impacts on the ecosystem, environment or health, may develop slowly and could take more than a decade before they could even be identified and thus their monitoring could begin. Similarly, salinity development under certain circumstances may take 15 to 20 years, but in other cases may take only 2 to 3 years depending on physical conditions, drainage provided and effectiveness of the management process. Thus, irrigated agriculture projects need regular monitoring and evaluation, even when the projects appear to be functioning efficiently for several years. The time dimension also makes inter-comparison of impacts of different irrigation projects a difficult task.

This chapter deals primarily with monitoring and evaluation of projects. There is, however, another important aspect, monitoring and evaluation of programmes. While conceptually the requirements for monitoring and evaluation of both projects and programmes may be similar, the clients are often different and the type of information, including level of aggregation, may have wide variations. Since different irrigated agriculture projects have different levels of impacts, at the programme level monitoring and evaluation has to be so designed in order that policy-makers get clear answers to questions like what aspects of a programme are doing especially well and poorly, what are the bottlenecks, and whether the designated beneficiaries are benefitting from the programme and, if so, to what extent? Such programme evaluation would generally cover many projects across a country, and thus it is necessary that the observers be trained to gather data in the same manner, using identical procedures and means. They should interview similar types of sources and pose similar questions. Without such compatibility, it is not possible to provide meaningful and credible information to policy-makers, who can then assess how effective the programme is and how it can be improved.

In terms of a rational monitoring and evaluation framework, irrigated agriculture projects can be categorized into four interrelated levels:

1. Planning, design and construction of physical facilities;
2. Operation and maintenance of irrigation and drainage facilities;
3. Agricultural production; and
4. Achievement of socio-economic objectives.

An important aspect to note is that these four levels are not sequential: they are generally simultaneous. For example, for large-scale irrigation projects, which often take 8 to 20 years for planning, design and construction phases, irrigation to new areas is introduced in sequence. Accordingly construction of some physical facilities may be continuing in one area, when another area may be receiving irrigation. Thus, the project will have all four levels of activities simultaneously. However, the first level is a discrete phase, which is completed as soon as the construction of physical facilities is over. In contrast, the rest of the three levels require regular monitoring and evaluation during the life of a project to ensure that the overall system is being managed at the desired efficiency, and that the objectives of the project are being continued to be met.

Planning, Design and Construction of Physical Facilities

Of the four levels, activities at this level are probably the easiest to handle methodologically in terms of monitoring and evaluation. This is also an area where some form of monitoring and evaluation has always been a standard practice. Engineers generally monitor progress in planning and design of the project; availability and use of equipment and construction materials; progress in the construction of various physical facilities and how they compare with the planned schedule; project expenditures at regular intervals; manpower availability and construction schedule.

Evaluation at this level has special significance for large-scale water development projects, where irrigation is generally developed in phases. On the basis of the evaluation of the performance of the first phase, it is possible to modify the proposed irrigation system of later phases, such as alignment of distributaries, canal lining, reduction or enlargement of irrigated area on the basis of water available, cropping pattern, water charges, and operating and management practices (Biswas 1987).

Generally, for most engineering and technical aspects, technical inspection and supervision and cost-accounting systems are normally integrated within a project in some fashion. What may be necessary is to review the existing or proposed system to see if some further improvements can be made to make the system more efficient.

There are some areas at this level, where monitoring is desirable or even essential, but is seldom performed except in an anecdotal fashion. Among these are employment generation and beneficiary participation.

Even though employment generation during the construction phase is an important benefit of any major irrigated agriculture project, and unemployment and under-employment are chronic problems in most developing countries, it is seldom monitored properly. If construction processes are specifically designed to use labour-intensive methods, employment generation can be maximized, and consequently, a large number of unskilled and landless labourers, including women, can benefit from them.

Beneficiary participation is another area which needs more attention. Very seldom do beneficiaries participate or are consulted on issues like canal alignment, which may be important in terms of equity distribution.

Operation and Maintenance of Irrigation and Drainage Facilities

Operation and maintenance (O&M) is one of the most under-estimated aspects of irrigation projects in developing countries. And yet, if the benefits from irrigation projects are to occur on time and to the specific target groups, it is essential that O&M be carried out efficiently to ensure that water supply is reliable, farmers at the tail end receive their regular and fair share of water and the drainage system is functioning properly so that salinity and water-logging problems do not occur. A review of past irrigation projects will indicate that most project agencies are generally not ready to undertake O&M work when the construction phase is completed. O&M still receives low priority, at least when judged by the actual performance, by both governments and donor agencies. Thus, not surprisingly, funds available for O&M are mostly inadequate or are not released on time. Often until a major crisis appears, mainte-nance efforts continue to be postponed. During this period, the efficiency of the project continues to decline, and during a crisis situation, generally the problem faced is more complex to resolve technically and more funds have to be extended than had the maintenance works been carried out on a regular basis.

Another issue worth noting is the fact that poor though O&M is for irrigation, it is generally even worse for drainage. Poor drainage contributes to salinity and waterlogging development, but since such problems take some time to develop, the magnitude and extent of the problems may not be realized until they become serious.

Agricultural Production

The fundamental objective of any irrigation project is to provide efficient and reliable water control in order to increase agricultural yields. Efficient water control, referred to at the previous level, by itself is not the sufficient condition to maximize agricultural production, which simultaneously requires other essential inputs such as seeds, fertilizers, pesticides, machinery, energy, as well as extension, credit and marketing facilities. It is equally important to ensure that irrigation water and the various inputs are available to the farmers on an integrated and timely basis. For monitoring and evaluation at this level, all these factors—with the exception of irrigation water which has already been considered in the previous level—need to be considered.

Information needs to be collected at critical times for each cropping season, which can then be used to provide better coordination between the different organizations responsible for the various inputs and services. At the end of a cropping season, an overall performance review of the season may be necessary, which may assist in the preparation of an integrated, and more improved, plan for the subsequent cropping season.

In most irrigation projects, monitoring and evaluation of agricultural production may require more continuing and regular effort than any of the other three levels.

Achievement of Socio-economic Objectives

The fundamental objective of any irrigation project is to increase agricultural production, which will not only increase availability of food for people, but also directly contribute to increased income generation of both farmers and non- farmers. Increased productivity and the consequent rise in farm income could go a long way to achieve the socio-economic objectives of the project.

It is, therefore, essential to monitor the impacts of the project on the proposed beneficiaries. For example, it is quite possible that an irrigation project may enhance the employment and income potential of landless labourers due to intensified agricultural activities. Equally, it could replace overall employment potential by undue emphasis on mechanization, which could make the life of landless labourers far worse than the pre-project level. Similarly, it may be possible that the income of small farmers and landless labourers increases significantly due to the project, thus making more equitable income distribution in the area. Alternatively, the benefits could accrue primarily to the large farmers at the cost of small ones, which would make income distribution even more skewed than ever before. Depending on specific irrigation projects, both alternatives have been observed in the past.

It is equally important to monitor the impact of increased income on some quality of life indicators. For example, is the increased income improving the quality of life of the people in the project area, e.g. better nutrition and literacy rate, improved housing and health services, provision of clean water and sanitation, etc., or is it being primarily used for conspicuous consumption, as has been observed in certain instances?

From a management viewpoint, it is desirable that monitoring and evaluation of the achievement of socio-economic objectives be carried out on a regular basis in order that a good picture is available on the status of the development. This will enable the managers to make appropriate policy modifications and interactions in time which will further maximize the benefits and simultaneously reverse undesirable trends. To this end, it is necessary to monitor both intended and unanticipated impacts.

It is not necessary to monitor the relevant socio-economic factors at frequent intervals. Some factors like education, health services or housing facilities can be monitored once every 3 to 6 years, depending on the situation and requirements.

Internal or External Evaluation

Basically there are two alternatives by which irrigation projects can be evaluated. These are internal, where a cell or unit is established within the project or the appropriate ministry, or external when the work is contracted out to a private company, research institute or a university. Both alternatives have their own benefits and costs, and the decision as to which alternative may work best is often project specific.

There are many benefits as well as costs for carrying out internal evaluation. Among the benefits could be the following:

- Staff members generally have a better knowledge and/or feel for what has really happened, the reasons as to why they happened, the actual constraints faced by the project management and feasible alternative solutions that are implementable.

- Evaluation can be handled less obstrusively since staff members of the evaluation unit may already have some rapport with the project staff, who may feel less threatened by such evaluations.
- Internal evaluators are likely to get more involvement of the users of the information as compared to external professionals, who may have less commitment to the project.
- It is likely to be less expensive than hiring high-priced external short-term consultants.
- The results of internal evaluation is more likely to be used since the evaluators may have more intimate knowledge of what could motivate the system to change and adopt new ideas.

Among the problems encountered during internal evaluation could be the following:

- Since no organization likes to have evaluations that reflect negatively on its programmes and operations, internal evaluators may face significant pressure to downplay negative findings and emphasize positive results. Evaluators may feel threatened if they are to question the decisions taken by the senior management. Organizations can influence internal evaluators by promotion control or even subtle threat of continued employment.
- There are many cases in national and international organizations where there is a greater stress to make their evaluation units visible than the quality or the use of the evaluation results. The evaluation unit becomes a symbolic representation of objectivity and quality of work of the organization. Such units thus serve a ritual function, which is more of a public relations exercise than to improve the management process.
- Internal evaluations may promote some vested interests.

Similarly external evaluations have their advantages and disadvantages. Among the advantages could be the following.

- Evaluations can be more objective and emotionally detached.
- They can provide new ideas, perspective and knowledge.
- They are likely to have more information on similar projects elsewhere and thus provide much-needed intercomparison.
- If external evaluating team contains people of stature and influence, the management is likely to pay more attention to the findings.

Among the disadvantages of external evaluations could be the following:

- External evaluators are not likely to be familiar with the structure, procedures and the working environment, which may make the evaluation process difficult, time-consuming and expensive. In certain instances, this unfamiliarity can even lead to wrong prognosis or conclusions.
- External evaluators may be more threatening to programme managers and thus they may withhold relevant information.
- Like internal evaluators, external evaluators can also be pressurized to downplay negative findings by threatening to withhold future contracts.

It is, however, possible to combine internal and external evaluations within an over-all monitoring and evaluation framework. A judiciously combined internal and external evaluation process can be more cost-effective than internal or external evaluation alone. Concern for the utilization of the results should be the driving force of any evaluation, and accordingly the decision as to whether to use internal or external or some mixture of the two should be decided on the basis of which is likely to produce the most usable results.

CONCLUSION

Monitoring and evaluation are integral components of any management of irrigation projects. However, it does not mean that if monitoring and evaluation are carried out, this will automatically improve the efficiency of management of projects. On the basis of a review of monitoring and evaluation activities of irrigated agriculture projects in Asia and North Africa, the general situation appears to be that monitoring and evaluation are having far less impact on the management process than expected or even possible.

One of the main reasons for this sad state of affairs is that monitoring and evaluation is being imposed from above by donor agencies, both multilateral and bilateral, on developing countries. A stipulated condition of any loan or grant by the World Bank, International Fund for Agricultural Development or similar funding agencies has been to establish monitoring and evaluation activities within a project. The monitoring and evaluation requirements of these external agencies are not uniform. The managers at the project level do not have a good understanding of the process. Under such unsatisfactory conditions, monitoring and evaluation gets done not because the national project managers feel that it is necessary, but essentially because it is a stipulated condition of the loan or grant. Accordingly, it is not surprising that monitoring and evaluation activities generally lack a sharp focus and the processes are seldom constructively reviewed either by national or international agencies. Monitoring and evaluation in many projects have become routine and perfunctory affairs that are done mainly because of administrative requirements, wherein activities and impacts are routinely monitored and documented, reports are neatly filed, but the project activities continue merrily on their way, unaffected in any sense with a 'business as usual' attitude.

For any evaluation to be used, it must be credible, which means the users must perceive it to be objective, accurate and fair. This means the evaluators must establish a basis of mutual trust before the results are delivered to the potential clients. Furthermore, evaluation should be fair to all groups that may have a stake in the results. Evaluation reports should be clear, unambiguous, balanced in terms of strengths and weaknesses, and contain justifiable conclusions and recommendations based on adequate and reliable data sources. Fuzzy and amorphous recommendations are likely to ensure that the results will not be used.

For monitoring and evaluation to succeed, we need a new ethos. As Brown (1976) has aptly noted, the heart of evaluation is an attitude, a frame of mind which enables us to review the project activities and performance in a constructively critical light. This should be done with emotional detachment. Managers need to develop a new evaluative mindset that allows them to appraise the performance, reflect on what has been learned for future activities and then adjust policies, if necessary, in response to what has been learned. Without such an ethos, it is unlikely that benefits of monitoring and evaluation can be fully harnessed.

REFERENCES

Biswas AK. 1987. 'Monitoring and evaluation of irrigated agriculture: A case study of Bhima Project, India', *Food Policy*, vol. 12, no. 1, pp. 47–61.
Brown DW. 1976. 'Evaluating development programs', *National Development*, October, pp. 32–40.

Lai KC, Felton MW. 1986. 'Estimating food production changes and project monitoring and evaluation in Nigeria', *Agricultural Administration*, vol. 22, pp. 161–73.

Sagardoy JA. 1985. 'Methodology for evaluation of small irrigation systems', National workshop on Evaluation of Small-scale Irrigation Project Performance, Bangkok, Thailand, 17–20 December 1985. June 1986. FAO Regional Office for Asia and the Pacific, Bangkok.

Steinberg DJ. 1983. 'Irrigation and AID's experience: A consideration based on evaluations', Agency for International Development, Washington DC.

CHAPTER 6

Management of Water Projects

WANCHAI GHOOPRASERT

Water projects cover a wide range of activities. Since water is considered a scarce resource, especially in arid and semi-arid countries,
and is essential for all forms of life, the efficiency in the use of water
must, therefore, be maximized. Furthermore, water resources
development should be multipurpose, e.g. drinking, irrigation,
industrial use, hydropower, flood control, navigation, recreation
and wildlife enhancement.

Maximization of water use efficiency requires careful planning
and good project management including proper budget, manpower
and equipment allocations, and on-time financial disbursements.
This chapter first deals with the general concept of planning, starting
from the macro or the national level, to the micro or the project
level, with emphasis on water projects. Then operation phase is
discussed, when good management is required for project success.

PLANNING AT VARIOUS LEVELS

Normally, planning starts at the macro, or the national level. At
this level, the planners have to decide the growth rate of and the
resulting investment ratio for the country as a whole. From this
ratio, a country can determine the extent of mobilization of the
goods and services possible for the investments that can be realistically made. Should it be necessary for goods and services to be
imported from abroad, foreign exchange will be required. If the
country does not have enough foreign exchange, it may have to
borrow and/or obtain grants to cover the deficit. Similarly, if it
does not have enough budgetary resources to procure domestic
goods and services, it generally has to borrow from the internal
capital market.

The next level of planning deals with the intermediate or the
sector level. During this phase, decisions must be made as to which
sectors of the economy investments are required. Since amount

available for investment is always less than what is needed, relative priorities of different sectors and areas need to be established. This decision is influenced by many factors, including politics. For example, even if the planners would like to spread economic development evenly over the country, they may still need to invest relatively more in certain geographical regions which are less developed in comparison to other areas. On the other hand, if the country has a growth-oriented policy it may want to invest in certain specific sectors that have low capital/output ratio so that output will grow fast. Or, the policy-makers may decide that certain sectors such as agriculture, irrigation or energy are priority areas which will receive significantly higher levels of investment.

A classification of various sectors of the economy is given in Fig. 1. It is likely that in a growth-oriented economy, more investments are channeled into the directly productive sectors. However, development in these sectors will demand investments in complementary infrastructure such as roads, power generation, water supply, housing for workers and so on. The end result is that one has to think in terms of total investment package instead of directly productive investments only.

Since the problems facing developing countries are immense, and different components of the sectors require varying degrees of resources and time for development, it generally will take several macro plans over a number of years to complete all the tasks satisfactorily. Gestation period of large-scale water development projects may extend to a decade or more, and total time necessary for construction of all hydraulic structures and supporting agricultural

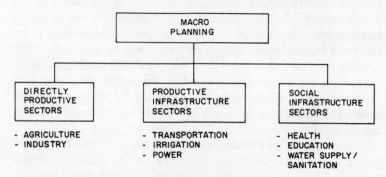

FIGURE 1. Relationship between macro planning and various sectors

development and associated infrastructure projects may require another decade. Equally, such projects may yield satisfactory results in the initial stages of operation, but in the long run, more sophisticated management may be required to ensure high degree of productivity and their overall sustainability. On the other hand, education is a slow process, where results may need more time to be visible compared to an irrigation project.

It is during the next planning phase, which is the micro level, that decisions are made to which specific investment projects should be executed within the chosen sectors and for regions. In practice, there may not be much of a choice in terms of projects to be executed, especially over the initial years of the investment planning since varying lead times are required to identify and prepare project documents on the basis of which informed decisions can be taken. Adequate expertise may not be available in many developing countries for project identification and preparation. Many of the investments could be predetermined, either because they are already under implementation, or because they are ready for execution.

The selection of individual investment projects is normally not handled by macro planners, but by those who are more familiar with and have more expertise on the sectors concerned and the projects. Apart from fitting well with the overall sectorial and regional priorities, each project must be economically justified, i.e. the discounted benefits should be higher than the discounted costs, and they should be environmentally sound.

PLANNING WATER PROJECTS

Figure 1 shows that water projects may practically fall in all of the sectors, namely, agriculture, industry, transportation, irrigation, power, health and water supply/sanitation. Also, even though their inclusion in the education sector may not be initially obvious, it is through the educational process that people can be trained to plan, design and manage water resource projects.

It is thus evident that it is impossible to combine all of the water-related activities into the organization since in that case the resulting organization will be so big that it cannot be properly managed. Accordingly, in all countries, the responsible organizations are divided under several ministries, which often means that the resulting coordination and collaboration between the concerned ministries on different aspects of water management leave much to be desired. It should be noted that many of the interactions between human activities and the water cycle are often not unplanned. Such interactions are not properly managed and hence they take place in an uncontrolled manner. Many countries are therefore taking an integrated approach to better manage their river basins.

A river basin may cover part of a country, the whole country, or several countries, in which case international relations between the countries concerned become an important factor for water development and management. For very large river basins, it may be necessary to divide them into smaller sub-basins for planning and management purposes.

An integrated approach to river basin planning and management, if carried out properly, can handle the cross-sectorial activities effectively. In order to illustrate this point further, the following examples are given to show the interactions between the management of the water cycle and the human activities.

Activities	Impact
Water supply, sewage disposal, irrigation and drainage	Provide direct intervention to the water cycle and have direct impacts on the environment.
Urban development	Increased storm runoff, higher demand for water, increased pollution load; changes in flood hydrograph patterns.
Construction of roads and railways	Increased storm runoff, higher pollution risks from possible accidents, fixing river courses at crossing points, may affect flood hydrograph and river morphology.
Agricultural development	Possible changes in runoff and sediment loads; higher pollution levels from fertilizers, pesticides and other farm activities.
Mining	Increased sediment runoff, higher pollution levels due to various contaminants; may increase base flows (mine dewatering) but lower groundwater levels; may provide flood attenuation.
Energy development	Dam construction and reservoir creation for hydropower development; changes in the aquatic environment and human settlements; requirement of large quantity of cooling water for thermal or nuclear power plants, and ensuing thermal pollution.

Individual activity or groups of activities may take place in river basin such as that depicted in Fig. 2. It is, therefore, often advantageous to have the management of all the important aspects of the water cycle within the river basin under one multi-functional authority. This will contribute to better coordination of the various activities, more efficient use of the available water, improved environmental conservation, and better implementation of national or regional development policies such as aid to depressed or underdeveloped areas, promotion of tourism, or reverse a trend towards urbanization. Unfortunately, it may not always be possible to establish such a multi-functional agency for river basin management.

If the establishment of multifunctional authority is not feasible, the government should clearly enunciate national policies or master plans for coordinated developments. Otherwise, the general tendency could be for the authorities to compete with each other over specific issues and availability of resources, and in the process get involved in the inter-ministerial and inter-departmental infighting due to conflict of interests. Simultaneously, there could be areas where actions are necessary but no authority is paying any attention.

It should also be noted that the national policies, or appropriate master plans may not be easily formulated, since they invariably will provide more benefits to some authorities and interest groups than others. Equally they would adversely affect others. The government, in the final analysis, must try to balance out these opposing forces. Who benefits from governmental policies or individual projects and who pays the costs are fundamental considerations for any efficient decision-making process. Thus, it is not only the total benefits that are important but also a clear understanding of the nature of the beneficiaries. For example, water supply and irrigation are competing demands and allocation must be made on the basis of overall values to the society. Domestic water supply has, in principle, a higher value than industrial or irrigation water in nearly all countries. In reality, only part of the domestic water supply is essential for satisfaction of the basic human needs. It could be effectively argued that the rest of the current domestic use has the same value as irrigation or industrial water. The relative values of irrigation or industrial water requirements are case specific and depend on the particular circumstances. To cut off irrigation water at a critical period of the crop cycle may mean losing the entire crop whereas cutting of water for a short time to

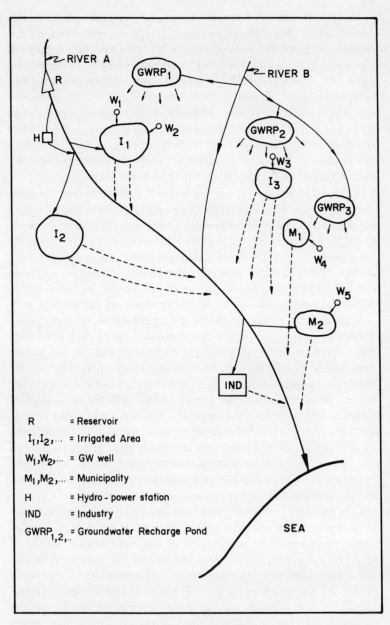

RIVER A

R

GWRP$_1$

W_1

RIVER B

H

I_1

W_2

GWRP$_2$

W_3

I_3

GWRP$_3$

I_2

M_1

W_4

W_5

M_2

IND

R   = Reservoir
$I_1, I_2, \ldots$ = Irrigated Area
$W_1, W_2, \ldots$ = GW well
$M_1, M_2, \ldots$ = Municipality
H   = Hydro - power station
IND   = Industry
GWRP$_{1,2,\ldots}$ = Groundwater Recharge Pond

SEA

FIGURE 2. A hypothetical river basin

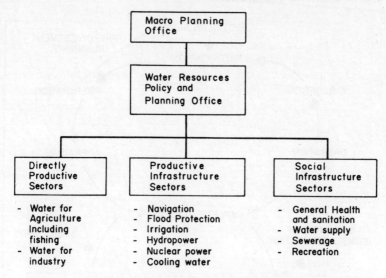

FIGURE 3. Coordination of water-related activities through water
resources policy and planning office

industry may mean losing only the production for the hours or
days for which water is cut off.

It should be clearly understood that water projects do not belong
exclusively to any one individual sector. Therefore, if there is no
effective coordination between the concerned sectors, the consequences could be very serious. A Water Resources Policy and Planning
Office could be established within the Planning Ministry to assure
proper coordination between all the agencies concerned, as shown
in Fig. 3. This office must, however, have appropriate executive
power and authority in order to ensure that the line agencies follow
their guidance.

THE PROJECT CYCLE

It is possible to maximize overall project benefits by following the
concept of project cycle. Through project cycle scarce resources
can be efficiently managed to achieve their development objectives.

The project cycle, shown in Fig. 4, starts with the identification
process and continues until evaluation. The important aspects of
each stage and their interlinkages will be briefly discussed.

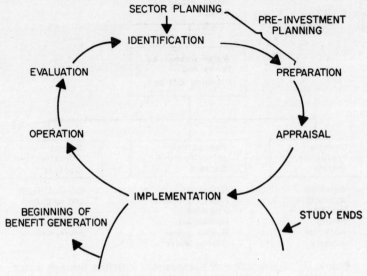

FIGURE 4. The project cycle

Project Identification or Formulation

During this stage, the objectives are clearly established. Potential problems and constraints are identified. Data are collected and analysed. Total costs, including investment and running costs, and benefits expected from the project are estimated. If the project receives the go ahead, recommendations for specific further studies are made.

Questions that planners should ask themselves when identifying projects are the following:

– Is the project technically feasible?
– Will total benefits exceed total costs?
– Who will benefit from the project?
– Is there a better alternative to achieve the same objectives?
– Will social and environmental cost of the projects be acceptable?

Project Preparation. At this stage, project objectives are more clearly defined, and its components are analysed in more detail. Additional data are collected and analysed to ensure that the objectives are achievable within the stipulated costs. The project is closely examined in terms of technical, economic, environmental, institutional and social feasibility so that risks and uncertainties are clearly defined and understood. Among the important issues that are to be considered during this stage include the following:

- clear definition of the objectives;
- establishment of specific project goals;
- study of the existing conditions;
- identification of problems;
- formulation of alternative solutions;
- study of the feasibilities of each alternatives; and
- recommendation of the best alternative for implementation.

Project Appraisal. This is the final verification that the project is, in fact, feasible and justifiable. Normally, it involves discussions among the implementing agency, other government and planning agencies concerned as well as the funding institution. Once the project is appraised and accepted, the study is taken as ended and it is now ready for implementation.

Project Implementation. This stage includes detailed design, procurement and construction of the appropriate facilities, purchase and installation of equipment, creation of new institutional arrangements if necessary and recruitment and training of staff. These activities are carried out in line with the alternative chosen in the preparation stage. Project supervision is very important for this stage. Careful monitoring of the progress and quality of construction, review of new information or unexpected problems are necessary to ensure the project is completed within the budget and time allocated.

Operation. The project begins at this stage to yield its expected benefits in the form of goods and services. It requires sufficient funds and their disbursements on time and adequate number of experienced staff to ensure that the project will continually yield maximum benefits over its planned lifetime.

Evaluation. Normally, evaluation should be carried out every 3 to 5 years after the operation begins. The main purpose is to assess to what extent the project is actually achieving its intended objectives and goals. Here, various problems are identified such as cost overruns, delay in constructions, technical malfunctions, or institutional weaknesses. Such information is essential for taking corrective actions so that the project becomes more successful; it can also be used as inputs when other new projects are identified.

Careful planning, ranging from project identification to appraised stage, carries a cost in terms of time and money. Normally, for water development projects, identification to appraisal takes 3 to 6 years to complete, involving considerable costs with no immediate benefits. Therefore, they may not be properly carried out. In reality, however, investment in these phases can yield very high returns by reducing potential future cost or cost over-runs as well as maximizing benefits.

IMPORTANCE OF BUDGET AND MANPOWER

As mentioned earlier, the project must be tested against the five feasibilities before implementation, namely, technical, economic, institutional, environmental and social. Only two of the five aspects will be discussed here to indicate how they influence the operation and maintenance of the project and, in turn, its overall success.

In the economic analysis, the initial and the recurrent costs are estimated. The recurrent cost includes the operation and maintenance costs. For example, in water supply projects, operation and maintenance costs consists primarily of personnel, chemicals, electricity, fuel and equipment. These costs are necessary to produce the expected benefits. It is obvious that without budget to employ personnel or to purchase appropriate chemicals, adequate quantity and quality of water cannot be made available to the consumers on a reliable basis. Unfortunately in developing countries, it happens quite frequently that budget is not adequately allocated first for construction of the facilities and later for their proper operation and maintenance. Also, budget could be mismanaged. Revenue collected from the water could be reallocated to some other activities and not for appropriate operation and maintenance of the project.

When a project is financed from an external source, be it a bilateral or a multilateral aid agency, only the capital cost is generally covered. It is up to the borrower to manage the project during the operational phase by ensuring that enough revenue can be generated to pay for the various expenditures. Provided there is no serious problem over repayment, it is during the evaluation phase of the project cycle when representatives of the funding agencies may visit the project to assess the achievement against the plan, and to identify problems and constraints when output is less than expected.

Institutional feasibility is also important to the success of a project. Institution includes not only the organization, and its structure, but also the manpower, proper training programme and the existence of and adherence to rules and regulations. The organizational structure, as well as the rules and regulations, could hinder or facilitate efficient operation. However, properly trained and experienced personnel at all levels are essential prerequisite for any efficient organization and in the operation of any project. It is impossible to run a large irrigation project without having adequate number of trained and knowledgeable professionals. Good manpower planning is important to any organization to ensure that the tasks entrusted to it are carried out correctly and on time.

Normally, economic and institutional feasibilities as well as technical, environmental and social feasibilities are appraised by the funding agencies concerned before the project is implemented. However, the executing agency plays the most important role by following the recommendations set forth in the appraised feasibility study report.

COMMUNITY INVOLVEMENT IN PROJECT MANAGEMENT

As mentioned earlier, community participation from the planning phase of the project to its operation and maintenance could play an important part in making the project a success. When dealing with small urban and·rural areas, socio-cultural factors seem to be more significant than large urban areas. These factors then are the basic requirements for a project to be successful. Water projects are no exception, since they should be managed as an integral part of the community and should not be isolated from the culture, or the way of life of the people. These factors can be grouped into three categories: community needs, community participation and technology transfer. Why and how these factors should be taken into account are discussed next.

Community Needs

It was observed that the failure of most community projects is usually because they did not satisfy the real or perceived needs of the communities. The following points are worth considering in evaluating the needs.

Demography. Data on population, including average family size, growth rate, morbidity as well as in and out migration patterns, should be accurately gathered, since they determine the quality of forecasts made for water demands.

Health situation. Major health problems, especially those related to water/sanitation should be identified.

Occupation. What are the major occupations of the residents and is there any seasonality of employment.

Level of interest. The community interest of having or improving water facilities should be compared to other potential projects. Commitments from the leaders should also be obtained.

Physical structure. Types and conditions of dwellings and the layout should be recorded. These kinds of information will help in determining the capacity and in designing the network.

Water use pattern and practices. Preferred sources of water and quantity used should be identified. Existing water facilities, if any, should be investigated and their popularity should be determined.

Not recognizing any one of those factors may cause failure. For example, for years the Thai authority invested hundreds of millions of baht in an attempt to provide villages in the northeast of the country with 'clean' deep-well water. To their surprise, the villagers never used this water for drinking as they considered it to be unhealthy to the extent of being poisonous. They would rather drink 'milky' water from shallow wells which they are used to, even though they have to walk many kilometres to fetch the water. If this problem could have been identified at an earliest stage, and appropriate steps were taken to overcome it, not only would the funds expended would have been used more efficiently but also the people concerned would have had access to better quality of water.

Community Participation

Once the community needs have been identified, the mechanism of fulfilling the needs have to be carefully considered. The above example from Thailand shows that if there was a need for water wells, community participation should have been a vital step which could have contributed to the success of the project. Only through such a process can a sense of belonging be developed in the residents. Often times when there is a minor problem with a well or a pump, it could be abandoned, since people feel that it is not their responsibility to repair it but of the government or the provider, which constructed it in the first place without consulting them. They would rather go back to their earlier unclean water sources than try to repair them. Had they been involved from the beginning, be it through their contribution of labour, money, or material, they might have felt that the well and the equipment belonged to them. They might have tried their best to maintain it. Usually, the following points have to be considered when planning for community participation:

Organization	– major local organization and their memberships, so that they can be properly consulted;
	– major local political or social factions which might affect participation; and
	– important characteristics that would determine the acceptablility and influence of outsiders.
Local technology and resources available	– local availability of construction materials; and
	– availability of skilled and unskilled labour.
Local attitudes, beliefs and associated practices	– general beliefs, important taboos, local habits, etc; and
	– general household cleanliness and public health practices.
Willingness and ability to pay	– household income;
	– household expenditure;
	– borrowing and savings customs;
	– ownership of land and house; and
	– availability of alternative water sources.

TECHNOLOGY TRANSFER

This step assures that the system installed will be properly operated and maintained. Quite frequently, complicated automatic systems are used which no one in the community really understands in terms of proper operation. Thus, minor repairs cannot often be made in case of breakdowns. It is essential that at least one operator is trained to operate and maintain the equipment properly. In addition, the following points should be considered.

Selection of Technology

The design should meet the needs of the users and yet be simple enough for local residents to understand. This will largely depend on the prevailing educational levels in the community and their general potential to be trained. This could vary from one community to another, and from one country to another.

Diffusion of Knowledge and Training

Usually this can be achieved through an appropriate organizational structure such as a selected committee, or cooperative, or even mass media. Selected community members may be trained in facility maintenance, keeping accounts, collecting fees, etc.

Motivation for Adoption

Periodic follow-up visits should be established to discuss any problems that could arise and also to motivate the community to care for the facilities. Feedbacks could also be obtained from these visits, which could be used in improving future projects.

Making water projects acceptable and operable by the users are not simple tasks. This is because the knowledge, values, beliefs and practices vary from place to place and usually depend on a multitude of factors like social class, ethnic group, educational background, traditions, etc. Moreover, projects in developing countries, especially in small urban and rural areas, greatly depend on cooperation from local leaders as well as from the residents. Though some techniques are available for incorporating people's participation in rural water supply projects, not much experience is available on how to effectively use them on large-scale water development projects. Current experience indicate mixed results for such participatory processes for large projects.

REFERENCES

Asian Development Bank. 1986. *Handbook on management of project implementation*, Manila, Philippines.
Canter LW. 1983. 'Impact study for dams and reservoirs', *Water Power and Dam Construction*, vol. 35, no. 7.
Ghooprasert W. 1986. 'Social aspects in water supply and sanitation', Proceedings, World Water '86 Conference, London, UK.
————. 1988. 'Planning for water supply in growing urban areas—A case example of Hat Yai/Songkhla, Thailand', Proceedings, International Symposium on Hydrological Process and Water Management in Urban Areas.
Ringskog K. 1979. *Pragmatic water planning*. The Economic Development Institute, The World Bank, Washington DC.
Sakawa M. 1979. 'Multi-objective analysis in water supply and management problems for a single river basin', *Water Supply and Management*, vol. 3, pp. 55–64.

CHAPTER 7

Watershed Management

ASIT K. BISWAS

INTRODUCTION

Even though watershed management has become an increasingly important aspect of water resources development in recent years in both developed and developing countries, it should be noted that its importance has been realized for at least some 2500 years. For example, Plato (428–348 BC) has graphically noted the impact of land use changes on river discharges (Biswas 1970). In his book *Critias*, Plato discusses the conditions of Athens some 9000 years before his time.

Furthermore it [the land of Attica in ancient times] enjoyed the fructifying rainfall sent year by year from Zeus; and this was not lost to it by flowing off into the sea, as nowadays because of denuded nature of the land. The land [then] had great depth of soil and gathered the water into itself and stored it up in the soil we now use for pottery clay, as though it were a sort of natural water-jar; it drew down into the natural hollow the water which it had absorbed from the high ground and so afforded in all districts of the country liberal sources of springs and rivers; and surviving evidence of the truth of this statement is afforded by all the extant shrines, built in spaces where springs did formally exist.

The importance of watershed management, however, has become especially relevant in recent years because of the increasing realization that the long-term sustainability of water projects is an essential requisite for human welfare. Large- and medium-scale water development projects are invariably capital-intensive, and economically they can be considered to be efficient only if the benefits accruing from such projects can be assured over their designed life periods. There is now considerable evidence from projects from many different parts of the world that the life spans of some reservoirs are going to be less than expected owing to more than anticipated rates of sedimentation, which may occur due to changing land use patterns or serious underestimation of sedimentation rates.

WATERSHED MANAGEMENT

Watershed management, in its broadest sense, can be considered to be an attempt to ensure that hydrological, soil and biotic regimes, on the basis of which water development projects have been planned, can be maintained or even enhanced; under no circumstances must they be allowed to deteriorate.

It is now well known that changing land use patterns affect water and soil regimes of any watershed. These changes, if properly planned, could be beneficial. However, in nearly all watersheds of developing countries, the changes are caused by unplanned and ad hoc activities. Individual activities that affect land use practices are generally small and incremental. However, when all the individual activities over an entire watershed during a specific time period are considered, their overall aggregated impacts could be substantial. Based on recent experiences, such aggregated impacts, unfortunately, tend to be deleterious in terms of watershed management.

The root cause of these changes is undoubtedly continual increase in human and animal populations. With rising human and animal populations, more and more land is required for shelter, food production, forage requirements and fuelwood. Since in most areas, good agricultural and pastoral land are already being utilized, people are forced to use marginal land, often in upper catchment areas having steep slopes. Forest is cleared so that the land can be used for agricultural or pastoral purposes.

Against a backgound of rising human and livestock populations, and the combined effects of agricultural encroachment and deforestation, land use patterns of most watersheds have undergone serious changes in recent years and are likely to be subjected to more changes in coming decades. These changes, for the most part, are unplanned. The extent of this increased pressure can be easily demonstrated by considering the case of the Sudan, where between the 20-year period of 1957–77, the human population increased more than six-fold, the number of cattle twentyone-fold, camels sixteen-fold, sheep twelve-fold and goats eight-fold (Biswas *et al.* 1987). Given such rapid increases in populations, it may not be possible to develop and implement watershed management policies which could be sustainable over a long term.

Much international attention has been focused in recent years on the present status of environmental degradation in the upland areas of various watersheds in countries like Nepal, India, Thailand, China or Ethiopia. What is, however, much less realized is that the environmental degradation of such areas is a main symptom of a disease but not the disease itself. The disease in this case is the economic conditions of the rural poor, who are forced to exploit the environment for their very survival. It is evident that no enduring solution to such an environmental problem is conceivable without relieving the pressures that are forcing the local population from a sustainable to an unsustainable relation to their natural environment. Until the pressing needs of the rural poor living in the watersheds are alleviated, the practice will continue, unless a very forceful policy of exclusion of humans and livestock from land can be implemented, which might satisfy the needs of sustainability of land, but certainly not of the people.

Currently, two main implications of improper or inadequate watershed management that are receiving increasing attention are sedimentation in reservoirs and changing patterns of streamflow. Reservoir sedimentation, by reducing storage capacity, is seriously damaging economic and long-term sustainability of many water projects. Streamflow patterns could change due to deforestation. While increasing sedimentation in many reservoirs all over the world is no longer a debatable fact, hydrological and environmental scientists are not unanimous in their views on the extent and nature of streamflow changes caused by changing land use patterns.

RESERVOIR SEDIMENTATION

Undoubtedly the most serious and widespread impact of watershed management, so far as water projects are concerned, is in terms of reservoir sedimentation. While the problem is global in nature, the magnitude of problem often varies from country to country, and even site to site within the same country.

All rivers carry sediments, but their concentrations vary from one river to another. Table 1 indicates the sediment yields of some of the important rivers from all over the world. The table shows magnitude of surface runoff per unit area, sediment yields in tons per km^2 of catchment area, and sediment concentration in ppm. It can be seen that sediment yields range from a high of 40, 500 ppm for the Haihe in China to a low of 34 ppm for the Zaire River.

Table 2 shows the total suspended solid and dissolved sediment loads for three African rivers (Martins 1984). It shows that the

TABLE 1. Sediment yields of selected rivers

River	Country	Catchment area (10^6 km^2)	Runoff (cm)	Sediment (t/km^2)	Yield (ppm)
Haihe	China	0.05	4	1,620	40,500
Huanghe	China	0.77	6	1,403	22,041
Chiang Jiang	China	1.94	46	246	531
Mekong	Viet Nam	0.79	59	203	340
Ganges/Brahmaputra	Bangladesh	1.48	66	1,128	1,720
Indus	Pakistan	0.97	25	454	1,849
Tigris/Euphrates	Iraq	1.05	4	50	1,152
Amur	USSR	1.85	18	28	160
Niger	Nigeria	1.21	16	33	208
Nile	Egypt	2.96	1	38	3,700
Zaire	Zaire	3.82	33	11	34
Mississippi	USA	3.27	18	107	602
Amazon	Brazil	6.15	102	146	143
Orinoco	Venezuela	0.99	111	212	191

Adapted from Milliman and Meade 1983; Mahmood n.d.

TABLE 2. Sediment rates of some African rivers

River	Catchment area (10⁶km²)	Average discharge (km³)	Dissolved solids		Suspended solids		Total land 10⁶t/yr	Per cent by solution
			10^6t/yr	t/km²/yr	10^6t/yr	t/km²/yr		
Niger (Mouth)	1.24	200	13.8	11.1	19.4	15.6	33.2	41.4
Zairo (Mouth)	3.70	1,330	43	11.6	44	11.9	87	49.4
Nile (Cairo)	2.96	89	21	7.2	122	42	143	14.6

(Source : Martins 1984)

TABLE 3. Reservoir sedimentation in China

Reservoir	River	Catchment area (km²)	Storage (10^6 m³)	Sedimentation		
				10^6 m³	Years	% of storage
Sanmenxia	Huanghe	688,421	9,700	3,391	7.5	35
Qingtongxia	Huanghe	285,000	627	527	5	84
Yanguoxia	Huanghe	182,800	220	150	4	68
Liujiaxia	Huanghe	172,000	5,720	522	8	11
Danjiangkou	Hanshui	95,217	16,000	625	15	4
Guanting	Yongdinghe	47,600	2,270	553	24	24
Hongshan	Laohe	24,486	2,560	440	15	17
Gangnan	Hutuohe	15,900	1,558	185	17	12
Xingqiao	Hongliuhe	1,327	200	156	14	71

suspended solid loads for all the three rivers exceed the dissolved solid loads, thus indicating the dominance of mechanical erosion in these watersheds. For the Nile, suspended solids carried accounts for 83.4 per cent of the total sediment load which is significantly higher than dissolved solids carried at only 14.6 per cent of the total. This is because of extensive wind erosion in the desert areas through which the Nile flows.

Three estimates are currently available on the aggregated amount of sediments carried to the oceans by the world's rivers. Strakhov (1967) estimated the total load to be 12.7 billion tons. Corresponding estimates by Milliman and Meade (1983) and Holeman (1968) are 13.5 and 18.3 billion tons respectively. In the absence of real data, it should be noted that these estimates are very rough: the real figure may be very different.

Construction of dams and reservoirs invariably changes the river regime. As the river approaches a reservoir, its velocity starts to drop. This reduces the sediment transportation capacity of water, which in turn increases the rate of sedimentation. When a river enters the reservoir, because of very low velocity, sediment deposition rates become very high. Accordingly, sedimentation is a normal process in all reservoirs. What good watershed management can do is to reduce the rate of sedimentation, and thus prolong the useful life of the reservoirs. Table 3 shows the extent of sedimentation in various reservoirs in China over specific time periods.

After the sediment loads carried by a river are deposited in a reservoir, water released from it is generally clear. Thus, the construction of the Hoover Dam has reduced the sediment discharges of the Colorado River at Yuma, Arizona, where it enters Mexico, from 135 million tons to only 0.1 million tons per year (Meade and Parker 1985). Similarly, the River Nile used to carry 100–150 million tons of suspended matter at Aswan before the construction of the High Aswan Dam (Biswas 1982), most of which is now deposited in the High Dam Lake.

Discharge of clear water from reservoirs, however, may create erosion problems downstrem. For example, in the case of the Aswan Dam, bank and bed erosion downstream in the Nile has become a problem. Even more serious has been the erosion of the Nile Delta, some 1000 km from the dam. Prior to the construction of the dam, the delta used to be built up during the flood season with the sediment carried by the river to the Mediterranean. This sedimentation compensated the erosion of the delta resulting from the ocean waves of the preceding winter. There was an equilibrium between sedimentation and erosion. With most of the sediments being trapped in the lake, enough sedimentation now does not occur in the delta, and this has resulted in serious coastal erosion in that area.

Forecasting Reservoir Sedimentation

Continued sedimentation over a period of years means that the storage capacity of any reservoir decreases with time. If the rate of sedimentation is equal to or less than the designed rate, cost-effectiveness of the project from the sedimentation point of view is not under question since the expected life of the reservoir would remain as planned or more. However, when the rate of sedimentation is higher than expected, economic life of a reservoir reduces, which means that the economic effectiveness of the project may become questionable. Unfortunately, for many reservoirs all over the world, the rate of sedimentation had been seriously underestimated for a variety of reasons.

First, sedimentation is a complex process, and the present state of knowledge for forecasting the rate of sedimentation leaves much to be desired. Much of the work is empirical in nature, and there is considerable doubt over its validity in different agro-ecological zones and under different socio-economic conditions.

While some of the early studies on sedimentation were carried out in India, many of the recent analyses are from Europe and North America. Analytical techniques developed on the basis of empirical work in the Western countries are often not applicable in various Asian and African countries for many reasons.

It should be noted that there are important differences between temperate climates where most recent erosion and sedimentation studies have been carried out and tropical climates (Biswas 1985), which makes technology and knowledge transfer from one to another a hazardous process. For example, rainfall and temperature distribution patterns in tropical climates accentuate the soil erosion problem. The yearly average rainfall between London in temperate climate and Sokoto on the southern border of the Sahel, do not differ appreciably: 568 mm and 668 mm respectively. However, when distribution of rainfall throughout the year is considered, the two cases are very dissimilar. The rainfall pattern of London, which is a temperate climate, can be characterized by a low but reasonably uniform monthly rate over the entire year. It varies from a maximum of 61 mm in October to a minimum of 35 mm in April. Similarly, rainfall retained in the soil is reasonably uniform. The situation is very different for Sokoto, where the rainfall is intense during July to September, but virtually non-existent between October to April. The rainfall varies from a maximum of 239 mm in August to zero between November to March. Furthermore, Sokoto has a significantly lower rainfall retention rate in the soil when compared to London. Thus, even though the total average annual rainfall in Sokoto is actually 15 per cent higher than in London, its distribution throughout the year is very uneven, making Sokoto very arid.

Rainfall has a direct impact on soil erosion all over the world, but the potential ability of the tropical rainstorms to cause soil erosion is far higher than in the temperate regions. This could be attributed to the high kinetic energy of the tropical rainstorms when compared to the gentler kinetic energy of rainfall in temperate regions. Kinetic energy of rainfall depends on the size of drops, intensity and wind velocity. While long-term detailed data on tropical rainfall are not available, it appears that median drop size of well above 3 mm are not uncommon. Drop sizes as high as 4.9 mm have been observed. From these data, a preliminary observation could be that the drop-size distribution of rainstorms is much higher in the tropics than in temperate regions.

Kinetic energy of rainfall is an important consideration since kinetic energy and the impact of raindrops initiates loosening and detachment of soil particles, the first essential step for soil erosion. Once soil is loosened, the particles are washed away, thus contributing to serious soil erosion problems.

So far as intensity of rainfall is concerned, it appears that its erosive power significantly increases at about 35 mm h^{-1}, which can be considered to be a threshold for erosion. Since more rainstorms in the tropics equal or exceed the level of this erosive threshold, the erosive potential is higher in the tropics when compared to the temperate parts (Hudson 1971).

Another climatic aspect further contributes to soil erosion. The rainfall and temperature distribution patterns in tropical climates, especially in those areas having pronounced dry and wet seasons, accentuates the problem of soil erosion. During the long dry season, there is some loss of topsoil due to wind erosion. However, far more damage is done during the onset of the rainy season. The vegetative cover, at the end of the dry season, is already reduced and often at an absolute minimum. Thus when a heavy thunder shower occurs, the water does not infiltrate into the soil as it might in light steady rain, and year after year soil erosion takes place due to surface runoff.

Another important factor in the tropics for predicting sediment yields is the instantaneous river discharge, and not discharge averaged over a period of time. It is not uncommon to find that one major flood in a river carries more sediment than the sum of all non-flood events during that year. Table 4 shows the maximum 5-day sediment loads observed in some Chinese rivers, which represent 33 to 75 per cent of the maximum annual sediment load.

TABLE 4. Sediment load at gauging stations of Huanghe and its tributaries in its middle reaches

River	Drainage area (km^2)	Maximum annual sediment load		Maximum 5-day sediment load	
		Year	Load in 10^6 tons	Load in 10^6 tons	% of total annual load
Huanghe	497,559	1967	2,460	814	33.2
Weihe	106,498	1964	1,060	400	37.8
Wuding	30,217	1959	440	185	42.2
Beiluo	25,154	1966	216	137	63.6
Kuye	8,645	1966	303	228	75.2
Huangfu	3,199	1959	171	97	56.8

Secondly, reliable hydrological, meteorological, geological and land use data for most watersheds in developing countries are not generally available. At best, such data may be available for limited periods and for localized areas. Under these conditions, it is a difficult task to predict accurately the rate of erosion and resulting sedimentation from the entire watershed.

Thirdly, all irrigation projects significantly change the prevailing land use patterns. Cropping intensities increase, cropping patterns are changed and fallow periods are reduced or even eliminated. These changes directly affect soil loss from the area. Table 5 indicates how the rates of runoff and soil loss change under different vegetative covers in India.

TABLE 5. Runoff and soil loss under different conditions in India

Location, soil type, slope, rainfall	Vegetative cover	Runoff as % of rainfall	Soil loss (t/ha)
Dehradun, silty clay loam, 9%, 1250 mm (3-year average)	Grass (*Cynodon dactylon*)	27.1	2.1
	Bare fallow	71.1	42.2
	Bare and ploughed fallow	59.6	155.9
	Natural grass	21.2	1.0
Vasad, alluvial soil, 2%, 791 mm	Natural fallow	2.1	2.0
	Grass (*D. annulatum*)	1.1	1.5
	Tobacco	26.0	2.3
Sholapur, medium black soil, 1.18%, 607 mm (9-year average)	Natural fallow	4.8	1.3
	Bare fallow	19.8	43.0
	Shallow cultivation	22.5	60.4
Manjri, deep black soil, 3%, 627 mm (5-year average)	Bare fallow (weeds removed by cutting)	23.9	54.3
	Cultivated fallow	25.0	87.5
	Shorghum (winter)	16.1	60.5
Dehradun, silty clay loam, 11%, 1117 mm (2-year average)	Grass (*Cymbopogon citratus*)	11.0	2.3
	Cultivated fallow	16.2	18.5
	Strawberry	26.6	23.1
	Pineapple	10.5	8.4
	Pomegranate	33.5	16.4

Adapted from Singh 1985

The magnitudes of runoff and soil loss from grasslands also depend very much on their conditions in terms of livestock grazing. Studies carried out at Deochanda, India, indicate that overgrazing not only reduces grass cover but also deteriorates topsoil by compaction and loosening of soil particles by animal hoofs. Soil losses and runoffs from overgrazed areas are significantly higher than ungrazed natural pastures. Results of the Deochanda studies are shown in Table 6.

TABLE 6. Runoff and soil loss from grasslands at different levels of grazing, Deochanda, India (Singh 1985)

Conditions	Runoff (%)	Soil loss (t/ha)
Natural pasture, no grazing	11	0.40
Proper grazing	19	0.79
Overgrazing	27	2.37

Because of the above-mentioned and other problems, forecasting reservoir sedimentation in developing countries has been a complex and difficult task under the best of circumstances. Table 7 shows the designed and observed sedimentation rates for 8 major reservoirs in India. It indicates that for these reservoirs, the average observed annual sedimentation rate is twice the designed value. It further shows that the average annual rate of loss in storage capacity due to sedimentation is 0.6 per cent of storage volume.

It should be noted that while concentrations of suspended sediments depend on river discharges, the relation between them is not necessarily one to one. Figure 1 shows the seasonal variations of discharges and total suspended solids for the Niger River for 1980–1. It shows that the peak sediment concentration was attained well before the maximum river discharge (Martins 1984). While sediment load is highest during the rainy season, it appears that large amounts of fine sediments are washed to the river as flow starts to increase during the onset of the rainy season. Peak floods may, however, carry substantially higher bed load. The second sediment peak in Fig. 1 could be due to atmospheric dust deposition from the surrounding desert.

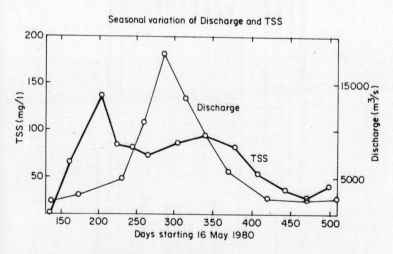

FIGURE 1. Seasonal variation of discharge and total suspended solids for the Niger River at Lokoja (source: Martins 1984)

TABLE 7. Design and observed sedimentation rates for selected Indian reservoirs

Reservoir	Year of impoundment	Storage capacity (million m³)	Annual sedimentation rate in ha-m/100 km²		Annual storage loss (% of volume)
			Design	Observed*	
Bhakra	1958	9870	4.29	5.93 (1979)	0.35
Gandhinagar	1960	7746	3.61	9.64 (1975–6)	0.29
Hirakud	1957	8100	2.52	6.82 (1982)	0.68
Maithon	1955	1357	9.05	12.38 (1979)	0.50
Mayurakshi	1955	616	3.75	16.48 (1969–70)	0.50
Nizamsagar	1956	456	2.38	6.37 (1967)	1.40
Panchet	1956	1497	6.67	10.00 (1974)	0.65
Tungabhadra	1953	3773	4.29	6.03 (1981)	0.45
Average			4.57	9.21	0.60

* Year of survey in brackets. Observed sedimentation rate is for the period between the year of impoundment and the year of last survey for which data are available

Economic Cost of Sedimentation

The economic cost of sedimentation in rivers and reservoirs is quite substantial. While no reliable estimate is available, Mahmood (n.d.) suggests that on a global basis, reservoirs are losing storage capacity at the rate of one per cent annually. Cumulatively it would mean a total annual water storage loss of 50 km^2, which Mahmood (n.d.) further estimates 'modestly' at $6 billion per year.

The one per cent estimate in annual storage capacity loss of reservoirs is likely to be an over-estimate for larger reservoirs. Analysis of 8 major Indian reservoirs shown in Table 7 indicates an average annual loss of 0.6 per cent. Similarly analysis of 19 reservoirs in central Europe, having storage capacities ranging from 1.5×10^5 m^3 to 23×10^6 m^3, indicated an annual storage depletion of 0.51 per cent (Cyberski 1973). According to Dendy *et al.* (1973), storage losses in the United States due to sedimentation decrease with increase in reservoir capacities, and amounts to about 0.16 per cent for capacities higher than 10^9 m^3. In contrast, average annual storage loss for small reservoirs having a capacity less than 10,000 m^3 is 3.5 per cent.

Sedimentation, especially in rivers, has another added economic cost in terms of dredging. It has been estimated that 376×10^6 m^3 of sediments are dredged every year from the water bodies of the United States at annual costs of more than $300 million (MR Biswas, 1979). In Bangladesh, its three major rivers—the Ganga, Brahmaputra and Meghna—carry an annual sediment load of some 2.4 billion tons, and Bangladesh Inland Water Transport Authority has to carry out every year nearly 0.8 million m^3 of maintenance dredging and 2 million m^3 of capital dredging at substantial cost (Biswas 1987).

CHANGES IN STREAMFLOW PATTERN

Changing land use patterns in a watershed affect streamflow. The magnitudes and durations of such variations depend on a variety of geo-hydrological, meteorological and agro-ecological factors. Because so many factors affect streamflow regimes, it is very difficult to determine the specific changes that may have occurred following alterations in land use practices in the watershed. Table 5 shows changes in runoff caused by differing vegetative covers under experimental conditions in India.

Probably the most-quoted example of changing streamflow

patterns due to alterations in upper watershed vegetations is that of Bangladesh. Writers of popular articles like Eckholm (1976) and Myers (1986) have made it a 'conventional' wisdom that deforestation and land use changes in the Himalayas in Nepal have contributed to higher flood runoffs and sediment loads during the monsoon season and lower dry-weather flow in rivers like the Ganga and Brahmaputra in Bangladesh. It has been further claimed that flooding in the Gangetic plain has systematically increased, both in terms of frequency and magnitude, during the past 50–100 years, and the devastating flood of 1988 in Bangladesh could be directly attributed to deforestations in the Nepalese Himalayas.

While this hypothesis has considerable public appeal and is quoted quite frequently in the present era of environmental consciousness, the following points are worth noting.

- There is very little reliable and replicable data which can prove or disprove the hypothesis categorically.
- Deforestation in Nepal was already well advanced by the middle of the eighteenth century, and in the Middle Mountains, which is the most densely populated region, all arable land had been converted by 1920–30. Little reduction in forest cover has occurred since 1930 (Ives 1989).
- Geologically, the Himalayas is a young mountain range, and has some of the highest erosion rates of the world.
- Preliminary studies by Gilmour et al. (1987) indicate that soil erosion and flooding are not necessarily due to deforestation, and are not likely to be significantly reduced by afforestation.
- High economic losses in recent years in the plains of Bangladesh can be attributed to a significant extent to rapid population growth, and consequent high level of economic activities. Thus, floods of a given magnitude will create a much higher level of damage at present than, say, 25 years ago.

It is not possible at the present state of knowledge to make cause-and-effect linkages between deforestation in the Himalayas and increased flooding and high sediment loads in rivers in Bangladesh. Urgent scientific studies are needed to determine whether flooding in the plain has actually increased in recent years, and if it has increased, to what extent deforestation in the Himalayas has contributed to this impact.

MANAGEMENT PRACTICES

The main emphasis of watershed management at present has been to reduce sediment generation in the catchment area so that storage losses of reservoirs can be kept to an acceptable level. To this extent, the main management alternatives have been to increase forest cover in the upper catchment areas, develop and implement appropriate land use policies and prevent overgrazing. In general, the results of such policies, even when they are carried out properly, take time to be visible. For example, if afforestation is practised, some improvement may be noted in about a decade or so.

One of the major difficulties of watershed management, which has not been addressed thus far, is to what should the cost of such practices be charged. The present tendency has been to charge the entire management cost to the water development project itself. This raises a fundamental question. Should afforestation costs be charged entirely to water projects, even though much of the deforestation damages may have occurred well before the project was designed? While benefits to water projects of afforestation are undeniable, society benefits from such practices in many other ways as well. If the entire cost of such afforestation is charged to a project, it may, in certain cases, no longer be economic and thus may never even be built. The real question thus is what percentage of afforestation costs are attributed to a water project and on what basis? The debate on such vital issues has not yet even started.

A second difficulty is institutional implications. Water projects are within the jurisdiction of Ministry of Water Resources, whereas forestry generally belongs to Ministry of Agriculture or Ministry of Environment. Very seldom has inter-ministerial co-ordination been effective in any developing country to ensure that rational watershed management practices are implemented in the early stages of the project. Because of the long time interval needed between the beginning of an afforestation process and some observable impacts, afforestation should be initiated during the planning phase of a project. One would, however, be hard pressed to identify even a single major water project where such practices were implemented systematically in the planning stage.

Check Dams

Even when afforestation has been practised in the upper catchment, sediments will still be generated by natural processes and by accelerated agricultural activities due to introduction of irrigation. One of the alternatives available for sediment and water control has been the use of check dams. Check dams are generally small. These low dams are built across gullies and streams to store flood runoff. The practice has successfully been used for many centuries in rural areas of several countries such as India, China, Sri Lanka, Mexico and the United States of America. Basically check dams provide upstream storage of flood waters, which can be used subsequently for irrigation and livestock.

Check dams could range from relatively simple structures built with stones, gravels and clay to fairly elaborate and sophisticated rockfill dams with concrete spillways. Many of the early check dams were simple structures that were built across narrow valleys, having somewhat impervious rock or soil strata. These dams required very minor changes in the local topography, and accordingly could be constructed relatively quickly with low financial investment as well as limited labour input. These dams not only controlled the flow of water, but also sediments carried by flood waters. With the reduction in flow velocities by the presence of these dams, the rates of soil erosion are also reduced. Construction of a series of check dams on a gully or stream can significantly reduce the overall rate of soil erosion. Furthermore, as the flow velocities are reduced, sediment present in flood water is deposited behind such dams.

As the silt deposited on the river bed increases every year, after a period of time a very fertile area is available for cultivation, especially when the stored water disappears. Thus, check dams are structures which can not only harvest seasonal flood waters but they also contribute to soil and moisture conservation.

While check dams have been used for centuries, their use in recent years for water and erosion control is receiving increasing attention in countries as diverse as China, Nepal, India and Ethiopia. In China, check dams have been very successfully used in many areas.

A good example of the use of check dams for water and erosion control can be found in the Jiuyuan Gully in Suide County, Shanxi Province, China. The gully is a small tributary of the Wuding River, which in turn is a tributary of the Huang-He (Yellow River). The length of the main channel of the gully is 18 km and it has a catchment area of 70.1 km^2, of which 2130 ha is used for agricultural production. Some 10,000 people live in this rural area and agriculture is the main source of livelihood (United Nations Environment Programme 1983).

Water and erosion control was the most serious problem facing the Jiuyuangou Peoples' Commune in the Gully catchment area. The catchment has a high gully density of 5.34 km per km^2. Before a control programme was initiated, the rate of annual soil erosion from the catchment was estimated at 1.27×10^6 metric tons, which was equivalent to an average soil loss of 18,116 metric tons per km^2. Because of the high silt content of the water in the gully catchment, it could not be used efficiently for irrigation.

The Commune initiated a combined programme of contour farming and check dams. By 1974, 727 ha of land had been provided with contour farming. In addition, 311 small and medium check dams were constructed, which primarily acted as silt traps, and 30 small reservoirs were built to store a total of 1.18×10^6 m^3 of water for irrigation. The general plan of the Jiuyuan Gully catchment is shown in Fig. 2.

The construction of this series of small- and medium-size check dams had a remarkable impact on the water use and erosion rates of the area. The irrigated area increased to 170 ha, which was eleven times the pre-construction period figure. The flood peak in the gully was 90 per cent less than before and the average annual soil loss decreased by 770,100 metric tons, which was a reduction of nearly 60 per cent. This meant that the total agricultural production in the area increased by 2.3 times within a period of only two decades.

Because check dams are small and widely dispersed over rural areas in many arid countries, it is not possible to comment on their overall efficiency. Furthermore, the check dams are often constructed by local people, based on past experience and broad rules of thumb. Thus, there are numerous types of such dams, based on different 'design' parameters, located in an immense variety of site-specific topographical and other physical conditions. Their maintenance often differs from one location to another. In addition, not even a single country has made a national survey of these dams. Under these conditions, only some general comments can be made on their advantages and limitations.

Check dams, when they are properly designed, constructed and maintained, can be considered to be a very useful small-scale alternative for water and erosion control in rural areas of arid and semi-arid countries. They are easy to design, construct and maintain.

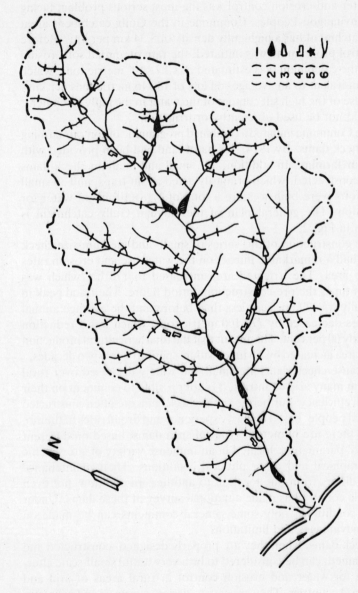

FIGURE 2. Use of check dams in the Jiuyuan Gully Catchment Area, China. Legend: (1) small check dams, (2) medium check dams, (3) small reservoirs, (4) village, (5) commune, and (6) streams

Labour and capital requirements are minimal, certainly significantly less than other sophisticated hydraulic structures. This means individual households or small communities can afford to build these dams without external assistance. Foreign exchange is generally not necessary, which could be an important consideration for many debt-laden developing countries at present. These simple structures can be constructed within a very short period of time, compared to large dams where the gestation periods are often more than a decade. Large-scale centralized institutions are not necessary for their construction, operation and maintenance. Also a series of such dams in an area can be developed incrementally.

Check dams have many limitations as well. They provide unreliable and discontinuous supplies of water, which means communities must have access to other alternative sources of water. Because of their decentralized nature, they often suffer from poor quality of design, improper construction and inadequate maintenance. Frequent repairs are necessary, but these repairs can be carried out quickly and within a limited cost. Many check dams are very vulnerable in terms of water quality contamination, and they often act as the main foci of water-borne diseases, especially those transmitted by mosquitoes.

CONCLUSION

Watershed management has become an important consideration for sustainability of water development projects. While the desirability of good watershed management is not in doubt, achieving it is not an easy task. It would require simultaneous achievement of many tasks, among which are afforestation, strict control of land use practices, and more emphasis on small-scale structures such as check dams for better soil and water conservation. Land use practices are generally very sensitive issues in most countries, at least politically, and thus to what extent it may be possible to develop and implement a rational land use policy is always difficult to predict. Furthermore, since watersheds of medium- and large-scale projects cover large areas, ensuring appropriate land use practices over entire catchment areas may not be a feasible process.

It should also be noted that afforestation of degraded lands needs substantial capital investment in early years, with no direct financial return to the investors. Thus, in most developing countries the governments would have to play the main role in afforestation, and/or subsidize private efforts substantially. Many governments are unlikely to give it priority at present in terms of investment because of their current economic conditions as well as financial requirements for other competing activities.

However, for most countries, watershed management should not be viewed from the narrow perspective of benefits to water projects alone: it should be considered to be essential for soil and water conservation, which in the long run will enhance the prospect of self-reliance of nations in terms of food, fibre and energy. Viewed in this holistic perspective, watershed management has to be considered to be a priority activity.

REFERENCES

Biswas AK. 1970. *History of hydrology*. North-Holland Publishing Co., Amsterdam, 336pp.
———. 1982. 'Environment and sustainable water development' in *Water for human consumption*, Cassell Tycooly, London, pp. 375–8.
———. 1984. *Climate and development*. Cassell Tycooly, London, 146pp.
———. 1987. 'Inland waterways for transportation of agricultural, industrial and energy products', *International Journal for Water Resources Development*, vol. 3, no.1, pp. 9–22.
Biswas AK, Masakhalia YFO, Odero-Ogwel LA, Pallangyo EP. 1987. 'Land use and farming systems in the Horn of Africa', *Land Use Policy*, vol. 4, no. 4, pp. 419–43.
Biswas MR. 1979. 'Agriculture and environment' in *Food, climate and man* (eds. MR Biswas, AK Biswas), John Wiley, New York.
Cyberski J. 1973. 'Accumulation of debris in water storage reservoirs in Central Europe' in *Man-made lakes: Their problems and environmental effects* (eds. WC Ackerman, GF White, EB Worthington), Geophysical Monograph 17, American Geophysical Union, Washington DC.
Dendy EF, *et al.* 1973. 'Reservoir sedimentation surveys in the United States' in *Man-made lakes: Their problems and environmental effects* (eds. WC Ackerman, GF White, EB Worthington), Geophysical Monograph 17, American Geophysical Union, Washington DC.
Eckholm EP. 1976. *Losing ground*, W.W. Norton & Co., New York, 223pp.
Gilmour DM, Bonell M, Cassells DS. 1987. 'The effects of forestation on soil hydraulic properties in the middle hills of Nepal: A preliminary assessment', *Mountain Research and Development*, vol. 7, no. 3, pp. 243–9.
Holeman JN. 1968. 'The sediment yield of major rivers of the world', *Water Resources Research*, vol. 4, no. 4.
Hudson G. 1971. *Soil conservation*. Batsford, London, 320pp.

Ives JD. 1989. 'Deforestation in the Himalayas', *Land Use Policy*, vol. 6, no. 3, pp. 187–93.

Mahmood K. (n.d. 1988?).'Reservoir sedimentation: Impact, extent, and mitigation', Technical Paper No. 71, World Bank, Washington, DC, 127pp.

Martins O. 1984. 'Rates of mechanical erosion in African Sawanna River', Proceedings, 4th Congress, Asia-Pacific Division, International Association for Hydraulic Research, Chiang Mai, Thailand, 11–13 September, vol. 1, Asian Institute of Technology, Bangkok, pp. 259–73.

Meade RH, Parker RS. 1985. 'Sediment in rivers of the United States', National Water Summary 1984, Water Supply Paper 2275, US Geological Survey 2275, Washington DC.

Milliman JD, Meade RH. 1983. 'World-wide delivery of river sediments to oceans', *Journal of Geology*, vol. 91, no. 1.

Myers N. 1986. 'Environmental repercussions of deforestation in the Himalayas', *Journal of World Forest Resource Management*, vol. 2, pp. 63–72.

Singh H. 1985. 'Status of grasses in soil conservation and integrated watershed management', Proceedings, National Seminar on Soil Conservation and Watershed Management, 17–18 September, New Delhi.

Strakhov NM. 1967. *Principles of lithogenesis*, Vol.1, (trans. JP Fitzsimmons). Consultants Bureau, New York.

United Nations Environment Programme. 1983. *Rain and stormwater harvesting in rural areas*. Cassell Tycooly, London, 238pp.

CHAPTER 8

Water Quality Management

———————◇———————

PART I

Surface Water Quality Management

N. C. THANH and D. M. TAM

WATER QUALITY CONSTITUENTS

Quality of water is affected by the presence of various substances. These substances can be generally classified under the following criteria.

Physical conditions, such as suspended solids which cause turbidity and prevent light penetration, sediments, and temperature;

Pathogens, including causative agents of water-borne diseases such as typhoid, paratyphoid, dysentery and viral hepatitis;

pH, and related constituents such as carbon dioxide, carbonate, bicarbonate, acidity and alkalinity;

Oxygen-consuming substances, which consist of organic compounds that are degraded by microorganisms using oxygen dissolved in the water. With large amounts of organic matter, oxygen consumption will exceed oxygen transfer from the atmosphere, and oxygen depletion in water will create profound impact;

Toxic substances, which include a variety of hazardous materials that endanger human health;

Substances of special ecological concern, like heavy metals or persistent pesticides, which are not readily biodegradable and accumulate in the food chains (the *bio-magnification* process);

Nutrients, like phosphorus or nitrogen, which promote the growth of algae and aquatic plants;

Chemicals creating aesthetic problems, such as phenol and sulphides, which have certain aesthetically objectionable properties;

Radioactive substances, which have harmful effects on the organisms in the ecosystems and consequently on human beings.

It should be noted that certain substances, at higher concentration levels, may have adverse effects on the quality while they may be beneficial at low concentrations. Thus, it may be misleading to condemn outright any constituent as a 'pollutant'. It may be preferable to consider any constituent to be suspected pollutant until its concentration is shown to be below a harmful level. Such constituents could be termed *potential pollutants* (Lamb 1985). The following sections will discuss those constituents that are the most significant potential pollutants.

Ions in Water

The major ions present in natural waters are:

– Cations (calcium, magnesium, sodium, and small amounts of potassium); and
– Anions (bicarbonate, sulphate, chloride, and small amounts of nitrate).

These ions come from the contact of the water with various mineral deposits.

Other natural ions which are less prevalent in water include:

– Cations (ammonium, arsenic, boron, copper, iron, and manganese); and
– Anions (bisulphate, bisulphite, carbonate, fluoride, hydroxide, phosphate, sulphide, and sulphite).

Most of these ions are also derived from mineral deposits.

In addition, some ions such as ammonium carbonate and sulphide are produced by bacterial and algal activities. Ions form micro-elements for nutrition of plants and animals and thus could be considered as being integral parts of the ecosystem.

The mineral deposits can also contribute to water quality in other ways. For example:

– iron and manganese could cause brown deposits and stains in water supplies;

- sulphate could be the cause of odours, corrosion of concrete sewers and, at certain levels, fish deaths;
- low levels (less than 1 mg/L) of fluoride in drinking water help prevent tooth decay, but higher levels cause tooth mottling, and very high levels may cause crippling; and
- high levels of sulphide released by decaying vegetation in a newly-formed reservoir may cause fish deaths, objectionable odours and corrosion.

Physical Conditions

Heat. Thermal and nuclear power plants use large amounts of water for cooling purposes. Where once-through cooling is used, the entire thermal load of the power plant is transferred to the receiving water body. Where closed-cycle cooling (such as cooling ponds or evaporative cooling towers) is used, the thermal load is transferred to the atmosphere on-site. The effects of thermal pollution depend on the capacity of the receiving stream, but most power plants raise the water temperature 8–11 °C in the once-through process.

Heat added to receiving water bodies as a result of cooling operations could have various impacts, both adverse and beneficial, viz.:

(1) imbalance of the ecosystem due to increased metabolic rates of the organisms living in water, or even death due to thermal shock;
(2) reduced solubility of oxygen in water and consequent DO reduction;
(3) reduction of the value of downstream water used as coolant;
(4) intensification of taste and odour problems; and
(5) improvement of the aquatic habitat in water, especially in temperate climates.

Impact (1) leads to the most far-reaching consequences in the aquatic ecosystems. Since biochemical reactions roughly double in rate with every increase of 10 °C, the rate of exertion of biochemical oxygen demand (BOD) is increased. This, coupled with impact (2), worsens the BOD-DO profile in the water body. As the temperature of the water rises, aquatic life is profoundly affected. Fish may die from thermal shock, reproduction may be impaired, or food supply reduced. Increased temperature also enhances the toxic effects of certain chemicals to fish. The thermal pollution may be particularly severe in the tropics and semi-tropical climates where the dry season stream flows are much lower than in the monsoon but the air temperature is at its highest.

On the other hand, increased temperature may have beneficial effects. Up to a certain level, the growth rates of fish and shellfish increase as the temperature increases, and the growing season is extended. Careful utilization of heated water for irrigation also helps increase crop yield and extend the growing season, particularly in temperate climates.

Pathogens

Water is an effective means for transporting pathogens, and in fact water-borne diseases are a major cause of deaths and sickness in developing countries. Up to 35 per cent of the potential productivity of many developing countries is lost because of these diseases.

Microbiological quality of water is often expressed by the concentration of indicator bacteria. Although total coliforms are customarily used as an indicator, especially in tropical climates, they do not possess adequate sanitary significance for several reasons. First, many types of coliform bacteria occur naturally in the soil, and very high populations of nutrient-rich natural waters may not be of sanitary significance. Secondly, under tropical conditions, coliforms tend to grow in natural waters and even during waste-water treatment before disinfection. In polluted warm waters, the numbers of coliforms can increase substantially and again may not have concomitant health implications. In temperate climates, natural bacteria have an optimum growth temperature of 15 to 20 °C which reflects the temperature of the local environment, whereas in tropical waters the temperature easily reaches 30 °C or higher. As many natural bacteria are able to grow at such temperatures, they would frequently interfere with total coliform tests which are conducted at 35 °C. Thus, relying on total coliform data under tropical conditions may lead to erroneous conclusions. Faecal coliforms serve better as an indicator of microbiological quality. Techniques of faecal coliform enumeration are similar to those for total coliforms, except that the incubation for faecal coliforms is 44.5 °C, a temperature which eliminates interference from any natural bacteria in the sample.

Besides faecal coliforms, faecal streptococci (typically *Strepto-*

coccus faecalis), clostridia (typically *Clostridium perfringens*), *Pseudomonas aeruginosa*, and some species of lactobacilli are occasionally used as indicators as they are always present in the intestinal tracts of humans. Unfortunately, all of these indicator bacteria also exist in animal faeces, so it is difficult to determine whether the faecal contamination is of human or animal origin. To solve this problem (at least partially), faecal streptococci (FS) are usually tested together with faecal coliform (FC). Since the ratio of FC/FS in human faeces is around 4.4 and in animal faeces less than 1, this ratio may be used to suggest whether faecal contamination is of human or animal origin. If the ratio is found in the range of 1 to 2, the conclusion is not clear. Knowing the source of contamination is a fundamental step in any system of water quality management. Although the ratio of FC/FS is not always reliable, in many cases it can be useful in identifying the possible source of contamination in runoffs. Geldreich (1976) and Wheater *et al.* (1979) have provided a review and evaluation of indicator systems used to distinguish sources of faecal pollution.

Toxic Agents

Trihalomethanes. Trihalomethanes (THMs) are formed when chemical elements of the halogen family (chlorine, bromine, and iodine) react with organic compounds. Main THMs of concern in water and wastewater treatment are chloroform, bromodichloromethane, chlorodibromomethane, and bromoform. THMs are suspected to cause cancer. Chlorine used in disinfecting treated water and wastewater can react with some organic compounds to form chlorinated hydrocarbons which can also cause cancer. For this reason, various alternatives to chlorine as drinking water disinfectants have been explored and applied (Drinking Water Research Division 1981).

Organic Compounds. Industrialization and technological developments have led to the increasing release into the environment of numerous organic compounds. For a large number of such compounds, very little is known about their precise effects on humans. For example, about 800 toxic compounds have been identified in the Great Lakes, the drinking source for about two-thirds of all Canadians and millions of Americans. Many of these compounds are known to cause cancer in animals. The best known chemical at present is perhaps dioxin, whose presence has been repeatedly found in minute amounts (less than 50 parts per 10^{15}) in various water supplies in Canada. Despite the fact that the Canadian interim health guideline for dioxin in drinking water is 0.15 part per 10^9, the very discovery of this very toxic man-made chemical has created much concern among the consumers. Recently, in Quebec province, dioxin has been found in mother's milk. However, it should be pointed out that during the last decade rapid advances in analytical methodology have substantially lowered the minimum detection limits of many compounds. This has led to increasing reports of the presence of toxic compounds in the environment, even though so little is known about their effects on human health and the environment.

Compounds of Ecological Concern

Heavy Metals. Most of the heavy metals present in water occur in ionic forms and owe their origin to human activity. They are toxic to microorganisms and ultimately to man. These heavy metal ions include arsenic, barium, cadmium, chromium, copper, lead, mercury, nickel, selenium, silver, and zinc. They come from a variety of sources: plumbing and gutters (cadmium, zinc), animal wastes (copper and zinc), paints (zinc), print ink (zinc), pesticides (mercury and zinc), and wastewaters from various industries. As they cannot be broken down, heavy metals accumulate in the food chains of the ecosystem. The process begins when low concentrations of heavy metals in water and sediment build up rapidly in aquatic plants and animals. The subsequent consumption of these organisms by other organisms in the food chains produce successively higher concentrations, until the concentrations in the organisms at the top of the food chains are sufficiently high to cause toxicity. The first massive heavy metal poisoning in modern history was caused by mercury which was released from a plastics plant into the Minamata Bay, Japan, in the 1950s. Local fishermen who regularly ate fish containing high levels of mercury, died or became disabled. Because of this incident, mercury poisoning is now often known as *Minimata disease*.

Pesticides. Pesticides include insecticides, fungicides, herbicides and algicides, although the term 'pesticides' is often understood as 'insecticides'. Common pesticides include:

- *Insecticides*
   Chlorinated hydrocarbons: aldrin, chlordane, DDT compounds, dieldrin, endrin, heptachlor, methoxychlor, lindane, toxaphene, hexachlorohexane (HCH)
   Organophosphates: diazinon, malathion, parathion
- *Herbicides*
   Carbamate: carbyl
   Chlorinated hydrocarbons: 2,4-D, 1,3,5-T, silvex
- *Fungicides*: copper sulphate, ferbam, ziram
- *Algicides*: mainly copper compounds.

Shortly after the introduction of synthetic organic insecticides, problems of water quality appeared. Many of the early insecticides, such as DDT, toxaphene and dieldrin, belong to the chlorinated hydrocarbon compounds. They are stable in the environment and, like heavy metals, accumulate in the food chains of the ecosystem. Molluscs, which feed by filtering large volumes of water, have been shown to contain DDT in concentrations millions of times higher than those of surrounding water. Organophosphate pesticides degrade more rapidly, but are more expensive. In the early 1950s research began to define the nature of water quality problems caused by agricultural pesticides. Runoff from agricultural land is a major source of low-level pesticide contamination of surface waters.

For many pesticides, concentrations in the sediment are higher than those in the overlying water. DDT is the most notorious pesticide. Because of its low cost and effectiveness against a variety of pests including disease-vector insects, DDT was used widely in the 1950s to 1960s, so much so that it was found in the Arctic where it had never been used. Fish-eating birds and birds of prey, being at the top of the food chains, suffered the most. DDT reduces the thickness of the shell of bird eggs, so making them break easily. The population of birds of prey in the USA reduced sharply and several species came to the verge of extinction.

Production of DDT up to the 1960s was about 100,000 tonnes per year, and its half-life in the environment may be as long as 20 years. Hence, despite the fact that DDT has been banned in many countries, a substantial amount will persist in the environment for many years. Meanwhile, there are signs that the ecosystems are recovering where there is no more input of DDT. The populations of various species of birds in the USA have increased significantly since DDT was banned.

Plant Nutrients

Plant nutrients elements in water exist in ionic forms. The two most important nutrients are nitrogen and phosphorus and problems caused by these are a result of human activities. The Green Revolution, with its high-yield crop varieties and intensive cultivation techniques, has led to substantial increases in the rates of fertilizer use. Consequently nutrient loads into natural waters have also increased due to the high levels in agricultural runoff.

Various forms of nitrogen exist in proteins, ammonia, nitrite, nitrate, urea, and nitrogen gas. Plants and some groups of bacteria synthesize proteins from inorganic forms of nitrogen (ammonia, nitrite and nitrate). Some bacteria (such as *Rhizobium* sp.) and plants (especially blue-green algae) are able to synthesize their protein from atmospheric nitrogen through a process known as *nitrogen fixation*. Bacteria then break down complex proteins into ammonia, nitrite and nitrate. This process is known as *nitrification*. Thus, monitoring the levels of different nitrogen forms in a water body of stream give some idea on the level of degradation of organic wastes released into the water.

Phosphorus exists at very low levels under normal natural conditions and is usually a limiting factor for bacterial and algal growth. Thus, when wastewater or agricultural runoff containing phosphorus is discharged into water bodies, algal blooms could occur. Eutrophication has occurred in water bodies receiving raw or treated household wastewaters since increased phosphorus concentrations accelerate the algal growth. Decaying algal mass decrease the dissolved oxygen (DO) in water. The eutrophication of the Great Lakes in the 1960s and of Lake Biwa, Japan, are good examples of the problems caused by phosphorus discharge into natural water bodies. In 1969, an effluent level of 1 mg/L of phosphorus, considered to be the lowest practical, was recommended in Canada. The Canadian Government adopted a two-pronged strategy using legislation and technology to reduce the amount of phosphorus in wastewater discharges (Schmidtke and Salloum 1985). Legislation was passed calling for a staged reduction in the phosphorus content of detergents to 20 per cent P_2O_5 by August 1970 and to 5 per cent by the end of 1972. The Canadian Federal Government and the Government of the Ontario then entered into an Agreement which called for the implementation and acceleration of pollution control programmes. Sewage collection and treatment systems were upgraded, phosphorus removal equipment installed, and research on phosphorus removal technology was given top priority. This integrated strategy yielded good results. The phosphorus concentration in raw sewage had increased steadily during 1967–70 from 8.3 mg/L to around 10 mg/L. Following the introduction of the first control regulation in August 1970, the raw phosphorus content in sewage decreased to 7 mg/L, and continued the downward trend to 5.7 mg/L in 1974. Within the year of 1972–3 the influent phosphorus loadings incrreased by 56 per cent.

Chemicals Causing Aesthetic Problems

In the early detergents, the extensive use of ABS (alkyl-benzene-sulphonate) during the 1950s and 1960s led to the deterioration of the aquatic environment due to the appearence in many places of persistent foam and suds in wells, dams and canals. ABS is highly branched molecule that is difficult for most bacteria to break down. In developed countries ABS was replaced in the early 1960s by linear-alkyl-sulphonate (LAS), which is more easily biodegradable. Phenols are other common chemicals that impart objectional tests and odours at very low levels. Some chemicals do not themselves cause aesthetic problems, but induce secondary reactions that cause problems. For example, sugars in effluents from paper mills sometimes stimulate the growth of filamentous bacteria *Sphaerotilus* in receiving water bodies. The chemical themselves do not impart any objectionable characteristics, but the filamentous bacteria may create difficulties in downstream water supplies by diminishing recreational and aesthetic values, and clogging fishing nets.

Radioactivity

Common radioactive elements include iodine-131, strontium-90, cesium-137, and radium-226. Natural waters may contain low levels radioactivity, especially those originating from deep underground. Radioactivity of human origin comes from the nuclear power industry, the use of radioactive isotopes in medicine and industry, and nuclear weapon tests. Radioactivity is of concern in waters used for drinking and food preparation, However, widespread sensitivity and awareness to radioactivity has meant that it is being subjected to control and monitoring perhaps more intense than for any other pollutants. There has been more justifiable concern about the accumulation of radioactivity in the food chains. Most waters contain radioactivity within WHO limits. Conventional water and waste-water treatment processes can remove radioactive elements, resulting in high levels of these elements in the sludges produced. These sludges should then be disposed of in a special way.

WATER QUALITY REQUIREMENTS
FOR DIFFERENT USES

Drinking Water Supply

The quality of raw water for drinking determines the type of water treatment and its cost. For a stream under consideration, two options could be compared based on the technological feasibility and costs: (i) permitting entirely or to some degree upstream pollution, then increasing the sophistication of downstream water treatment, or (ii) reduction of upstream pollution by wastewater treatment so that downstream water treatment is made simple. Developing countries sometimes choose the first option because it is easier to increase the sophistication of water treatment (which belongs to the public sector) than to enforce pollution control (the private sector). This option also reflects the reality of water pollution in that nature's assimilative capacity is exploited to the maximum. However it could endanger the health of those people who do not have access to the public water supply and depend directly on the polluted stream as their primary water source. In large urban centres, where water supply is an important government responsibility, with few point-source polluters (making pollution control at source easier), the second option is more logical.

With regard to treated water for drinking purposes, current guidelines on quality have been laid down by the WHO (1984), and these incorporate the latest information on aesthetic, inorganic chemical, organic chemical and bacteriological aspects. The WHO *guidelines* are simply targets to aim at and not rigid rules. For instance, the guidelines determine a zero level of faecal coliforms as adequate bacteriological quality. In developing countries, this condition is difficult to fulfil throughout, and therefore it could be gradually achieved by staged improvement of the local water supplies corresponding to the available capabilities and resources.

Quality and quantity should also be considered together. Where water is scarce and/or takes effort to fetch, the convenience of having a readily available water supply is appreciated.

Water Quality for Agricultural Production

The quality of water for agricultural use is related to its effects on plant growth. Plants may be affected directly by either high osmotic conditions in the plant tissues, or the presence of a phytotoxic constituent in the water. The presence of sediment, pesticides, or pathogens in irrigation water may not affect plant growth but may affect the acceptability of the product. Various elements accumulating in plant tissues may not affect plant growth but may be toxic to human or animal consumers.

Plant growth may be indirectly affected by the influence of water quality on soil. For example, the adsorption of sodium in water on soil particles will result in a dispersion of the clay fraction. This reduces soil permeability and results in surface crust formation which, in turn, hampers germination. Soils irrigated with highly saline water tend to flocculate and have high rates of infiltration. Thus, the levels and presence of ions are important parameters, because they affect not only plant nutrition but also the physical characteristics of the soil.

Water development projects may cause problems of salinization and alkalization. The effects of salinity, often directly measured by total dissolved solids (TDS), on the osmotic pressure of the soil solution is one of the major water quality parameters for cropping. Salinity is measured by determining the electrical conductivity (EC) of a solution. This measurement relates to the ability of salts in solution to conduct electricity. General guidelines of TDS and EC for uses of saline irrigation water are given below:

TDS (mg/L)	EC (mmhos/cm)	Crop response
<500	<0.75	No harmful effects
500–1,000	0.75–1.50	Detrimental effects on sensitive crops
1,000–2,000	1.50–3.00	Adverse effects possible on many crops, careful management practices (such as good drainage and sufficient water for leaching) required
2,000–5,000	3.00–7.50	Only salt-tolerant plants can be grown on permeable soil, with careful management practices

The sodium adsorption ratio (SAR) is often used to estimate water permeability. It is based on the relationship among sodium, calcium and magnesium—the most prevalent cations in soil. Sodium causes clay soils to disperse, resulting in reduction in pore size and decreased water permeability. Thus, high values of SAR indicate less permeability as calcium and magnesium are more tightly adhered to clay surfaces. Values for SAR are often considered together with values for electrical conductivity (EC).

Water Quality for Aquatic Life

In developing countries, fish constitute an important source of low-cost and readily available protein. Any planning for water resources development projects should consider the potential impacts of water quality on aquatic life. Water quality parameter of importance for healthy aquatic life include temperature, DO, turbidity, suspended solids, pH, and toxic materials such as hydrogen sulphide, ammonia, chlorine, phenols, pesticides, and heavy metals, notably zinc, copper and cadmium. Organic discharge up to a certain level into natural waters is beneficial. In a balanced aquatic ecosystem, symbiosis exists among the organisms. Heterotrophic bacteria in water consume organic matter and release carbon dioxide and simple nutrient elements, which are taken up by algae for reproduction and to produce oxygen. All aquatic animals benefit from oxygen produced by algae. In temperate climates, low DO in water may kill fish of high economic/recreational values (game fish) while low value fish (coarse fish) can survive. Eventually the polluted water bodies may contain only undesirable fish. In tropical climates, similar situations may happen but not severely since many species of carnivorous fish such as catfish and snakehead (*Channa straita*), and herbivorous fish such as Tilapia and grass carp (*Ctenopharygodon idella*) can withstand low DO levels by breathing at the water surface.

An imbalance will occur when the algae population increases to such an extent that the carbon dioxide released by bacteria is not sufficient for their growth. When this happens, many algal species will obtain the carbon dioxide needed by splitting bicarbonate ions and releasing carbonate ions which increases pH. In organic-rich algal ponds, pH in the water could be as high as 10–12, which is toxic to most fish. Too high an algal population may lower DO during the night, when photosynthesis (which consumes carbon dioxide and produces oxygen) stops, while respiration (which is the reverse of photosynthesis) continues as usual. This lack of dissolved oxygen can be clearly seen in the morning when fish surface to breathe from atmospheric air. Excessive algal populations may harm fish in another way. The gills of the Tilapia fish are clogged by algal cells, causing retardation in growth.

The detrimental effects of super-saturation of gases in water (such as nitrogen and carbon dioxide) to fish have only recently received attention. Super-saturation occurs when, for example, water containing a relatively high concentration of the chemical flows at a high rate (such as when flowing down a spillway). For sensitive fish, 102 to 103 per cent of saturation may start gas-bubble disease, and 115 to 125 per cent of saturation is lethal for most fish. Gas-bubble disease is characterized by the formation of fine bubbles on the fins and body surface of the fish especially the head, and the presence of gas bubbles in the vascular system. This causes blockage of the capillaries, leading to anoxia (lack of oxygen in tissues). When the bubbles break, the surrounding tissue is damaged, and becomes susceptible to bacterial infections. With too many bubbles in the blood vessels, the heart may lose suction, resulting in death.

Water Quality for Aesthetic and Recreational Uses

The common tendency in developing countries is for less stress to be laid on the aesthetic and recreational importance of water quality as other uses have a higher priority. This should not be the case in areas of high tourism, such as man-made reservoirs or river systems flowing to beach resorts. The aesthetic appearence of water bodies is perceived mainly through vision and smell. This means that the water should at least be free from floating objects or suspended solids and objectionable colours such as those given by industrial wastewaters. The water should also be odour free and have a reasonably low turbidity.

For recreational waters, the water should ideally be safe for body contact even when it is not intended for swimming. The reason is that, for many recreational uses such as fishing or boating, body contact with water is unavoidable and cannot be controlled by law. The quality constituents of concern are aesthetic (turbidity, colour, smell, floating/suspended objects), irritant chemicals, toxic compounds, and pathogens.

SOURCES OF WATER POLLUTION

Sources of water pollution include two broad categories, namely point and non-point. Point sources consist of municipal and industrial wastewaters, storm sewer outfalls, and any other sources which join water bodies through pipes and channels. Non-point sources include runoffs, atmospheric precipitations, and other diffused sources. The categorization is mostly useful in dealing with regulatory issues. Point sources can usually be quantified and controlled before discharge, whereas non-point are more difficult to manage.

Problems caused by non-point sources are:

– *Soil erosion and sediment transport.* Can result in habitat alteration and adverse effects on aquatic lives, filling of water courses and reservoirs, increase in cost and difficulty of water treatment, and reduction of recreational quality of water bodies.
– *Nutrients.* Nitrogen and phosphorus released from fertilizers and animal wastes stimulate plant bacterial growth, leading to eutrophication.
– *Heavy metals*, such as zinc, copper and mercury that are used in pesticides, paints, plumbing, and many other products.
– *Toxic chemicals*, mainly pesticides.

Irrigation-return flows may contain pollutants at concentrations higher than those in waters applied to the land because evaporation removes about 50 per cent of water, thus roughly doubling the pollutant concentrations. Runoffs from animal farms may contain high animal wastes, which contribute substantial amounts of oxygen-consuming substances and suspended solids.

Wastewaters

Municipal, agro-industrial and industrial wastewaters are important point sources of pollution. Water-borne pathogens are the main concern in municipal wastewaters, organic loads in agro-industrial wastewaters (Schaller and Bailey 1983) and organic and toxic pollutants in industrial wastewaters (Lund 1971) for industrial wastes. In developing countries, agro-industrial wastes also constitute significant pollution loads to the local environment. Examples are wastes from the processing of palm oil in Malaysia and Thailand, rubber in Malaysia, coconut oil in the Philippines and Thailand, tapioca starch in Thailand, and coffee in Kenya.

Another source of pollution that is less often recognised is urban storm runoff, which can contribute a wide variety of pollutant. In some cases, the pollution load of urban storm runoff may be even higher than that of the untreated sewage from the same community. Where sewage has been effectively treated, urban storm runoff will be a major source of pollutants and will have a strong impact on the receiving bodies.

Man-Made Reservoirs

Man-made reservoirs indirectly contribute pollutants through several mechanisms. Low turbidity due to sedimentation and high nutrients levels caused by the decay of submerged vegetation matter promote the growth of algae. This results in objectional odours, tastes, or colours. Stratification of deep reservoirs leads to anaerobic conditions which favour the dissolution of iron and manganese and production of sulphides. The discharge of impounded water in the lower layers may reduce the DO and temperature of the downstream areas far below than levels that would have been in the former free-flowing stream.

Other Sources

In developing countries, uncontrolled, open dumping sites are significant sources of organic matter (in particular refractory substances), heavy metals, and other toxic compounds, especially where hazardous materials are disposed of together with common garbage. The leachate from the dumping site will first pollute the groundwater and then surface water via ground outfalls. The residues from water treatment plants are also a threat to the environment since they contain bacteria, solids and organic material. In developed countries, correct treatment and disposal of these residues has become compulsory by law.

WATER QUALITY OBJECTIVES, CRITERIA AND STANDARDS

Water Quality Objectives

These refer to general aims or goals to be obtained usually within a fixed time frame. For example, an objective could be that within ten years no streams and rivers in a country should have anaerobic conditions as indicated by bad smells and a black colour. In most urban centres of developing countries rivers and canals have already been rendered anaerobic through discharges of untreated municipal sewage and industrial wastewaters, so it is questionable if this objective can be achieved. However, it is an express objective of the decision-makers to achieve as much as possible. Water quality objectives usually do not specify the means by which they could be achieved. As such, they may not bear any legal enforcement power. They simply indicate the overall objective towards which the regulatory programme is directed and defined.

Water Quality Criteria

The criteria represent the characteristics of water that are necessary or desirable for specific uses. Criteria are expressed in specific numerical forms and are thus more specific than objectives. A comparison of data on the quality of water under consideration with established criteria will indicate whether the water is suitable for its intended use and, if not, what changes in quality would be necessary to make it suitable. For example, one of the criteria for irrigation water is that the boron content should be less than 0.75 mg/L for sensitive crops. This figure should not be viewed as a rigid and absolute limit, because the effects of boron vary with the type of crop and with other conditions such as temperature, water pH, and the presence of other chemicals. While boron contents lower than 0.75 mg/L indicate better quality, higher contents do not necessarily exclude the use of water in irrigation but only represents a reduction in the margin of safety.

Water Quality Guidelines

These are numerical concentrations or general statements recommended to support and maintain a designated water use. Guidelines are somewhat more detailed and forceful than criteria, meaning they indicate a strong desire for achievement and compliance. Still, guidelines are not rigid and do not have enforcement power, but they clearly express the ultimate desirable quality that should be achieved. The World Health Organization in 1984 published the International Drinking Water *Guidelines* to replace their 1970 European *Standards* and 1971 International *Standards*, apparently to better reflect this philosophy.

8

Water Quality Standards

These standards form the yardstick by which regulatory agencies define their requirements for water streams and water bodies. By contrast with criteria or objectives, standards are rigid reflections of laws and regulations. They are usually viewed as absolute: either they are met or violated. It should be noted that not every country by all means should try to establish their own standards. For example, even a developed country like Canada has only *criteria*, not standards, for natural waters and *guidelines*, not standards, for drinking water. Only the Province of Quebec has enforceable drinking water legislation (Tam *et al.* 1986).

There are three basic water quality standards that can be imposed and enforced.

Technology-based Standards. These are guided, in the USA for example, by the concept of Best Available Technology (BAT) or Best Practical Technology (BPT), with an assumption that the application of the best technology is adequate for environmental protection. Sometimes technology-based standards are also known as *removal standards*, when they specify the degree of removal to be attained in the treatment plant.

Effluent Standards. These refer to the minimum concentrations of potential pollutants measured at the discharge points where the wastewater stream begins its contact with the environment. Effluent standards expressed in concentrations disregard the status of the environment: whether the receiving stream in a large river or a small spring, a wastewater treatment plant still has to ensure that its effluent discharge achieve the determined concentration limits. To correct this drawback, standards based on the mass of pollutants (for example kg of BOD per day) have been established and are now commonly used in the USA.

Receiving-stream Standards. These make use of the natural purification (assimilative) capacity of the receiving stream. With the same standard, a stream which has a higher assimilative capacity can receive higher pollutants loads. An example of receiving-stream standard in developed countries is a minimum DO level of 4.0 mg/L to be obtained at all sections of the stream.

For many people, technology-based standards imply that the pollution control technology itself is a limiting factor, and standards reflect this limitation. Like effluent standards, technology-based standards disregard the capacity of the environment to absorb a certain amount of pollutants. For instance, it has been found that the COD level in palm oil wastewater can be reduced at best only to 1000 mg/L by conventional treatment processes, and there is no viable means to remove the deep brown colour in the wastewater. In other words, technology exists to bring the COD level down further or to remove the colour, but it is not practical. If this waste treatment technology is accepted as standard, the environment around palm oil factories would still be heavily loaded with refractory materials and deep brown colour. In addition to waste treatment other measures, such as waste control at source or process modification, should be explored. On the other hand, requiring the best technology in developing countries may have serious implications. New technology may require new equipment and new skills to achieve the expected performance and, if these are not available, second best or even third best technology is preferable. Equally, the cost of the best technology may be prohibitive, and marginal benefits may not justify it.

Obviously, in many cases the environment may not benefit from technology-based or effluent standards. Under different conditions environmental over-protection could result, leading to unjustifiable costs. This is the major drawback of both standards. Effluent standards based on concentrations are unfair tò small-scale polluters since these people have to adopt the same sophisticated technology to produce the same level of effluent quality even though they release only small amounts of pollutants into the environment. Effluents standards based on pollutant mass may have to be 'tailor-made' for each specific water body, and so are quite complicated in terms of application. Despite these limitations, both effluent and technology-based standards still prevail since both are developed by engineers, chemists and biologists who are familiar with pollution control technology and analytical methodology. They are also easy to administer and apply, especially on a national scale.

Stream standards are the most rational and meaningful. To politicians and decision-makers, they are easier to understand. For example, 4 mg/L of oxygen means a healthy condition, and zero oxygen means a black colour and foul odours. Such a standard should serve as an ultimate goal for any water quality management programme. Some points, however, need to be noted when receiving-stream standards are considered.

– Standards lack precision and are highly site-specific. This is especially true when standards used in the West are imposed on a developing country. Water quality standards cannot always be imported from one country to another. Rather, each country should try to develop its own standards based on specific local conditions.

   The site-specific nature indicates even in the same country, there may be more than one set of standards. This concept of 'variable standards' is useful for each area of the country to deal with its own specific water quality problems.

– Standards should balance on the one hand long-term environmental impacts and on the other the economic technological and institutional capabilities of the country to monitor and implement. To do this, standards may need to be developed in stages according to the country's pace of development, but the initial level must take into account immediate costs and benefits with a clear understanding of the environmental impacts.

– Standards must be enforceable, otherwise their usefulness is limited. Unfortunately, in most developing countries this is usually the case. Any country can establish comprehensive standards, but most are not in a position to enforce them.

– Often, the common practice is to set unrealistically strict standards to provide for a large safety margin and to avoid institutional responsiblity. When compliance is impossible and the standards are violated then enforcement is quietly relaxed. However, such a practice easily leads to the loss of credibility and to an atmosphere of complacency, all of which impede the seriousness of future efforts to enforce environmental regulations.

WATER QUALITY MONITORING

Water quality monitoring is a far more complex task than water quantity monitoring because there are a higher number of parameters involved, higher costs in sampling and laboratory analysis, and higher requirements for skills and equipment. Whereas the same few water quantity parameters are measured at every station of the monitoring network, a higher number of water quality parameters must be selected specifically for each station. Equally water quality parameters to be monitored could vary with time. More equipment and skills are required to ensure that water samples are taken correctly, stored and preserved properly, analyzed reliably, and the results are interpreted properly. The site-specificity and the sophistication of water quality monitoring necessitates careful planning, appropriate implementation of the monitoring programme, and regular review.

Almost every developing country has its own water quantity monitoring programme which consists of measurements of physical parameters like rainfall and river discharge. Water quality monitoring and ecological surveillance including bioassays are less prevalent. Generally there is a missing link in these programmes, that is, what should be done next with all of these data. Commonly, large amounts of data lie on the shelves to gather dust. After substantial resources have been spent in gathering data with an elaborate station network using sophisticated equipment, developing countries often do not use the data collected. Thus, instead of collecting a lot of data, it could be more effective and efficient to focus the monitoring programme only on critical areas, where they are needed. An integrated plan with all necessary stages should be drawn up, the objectives of each stage defined clearly, and resource allocated appropriately to all the necessary stages.

Water quality monitoring must serve an explicit purpose, be it for environmental impact assessment, project planning and design, project evaluation, or for regulatory requirements. Since resource is always a constraint, only the most meaningful parameters should be selected for inclusion in the monitoring programme. For example, total coliforms need to be monitored together with faecal coliforms; or, once total nitrogen and ammonia have been determined, nitrate and nitrite could be lumped together by subtracting ammonia from total nitrogen, since determining nitrite and nitrate separately costs more but is not very useful.

WATER QUALITY INDICES

Environmental Indices

Due to the number of water quality parameters and the high variations of these parameters, a water quality index may be of help to describe the overall situation. Unfortunately, the number of water quality indices used in different countries (or even in different states in the USA only) is as high as the number of parameters themselves. Generally there are six basic uses of environmental indices (Ott 1978).

- *Scientific research*: for reducing a large quantity of data to a form that gives better insights into the prevailing environmental conditions;
- *Public information*: providing information to the public about environmental conditions;
- *Ranking of locations*: intercomparison of environmental conditions at different locations or geographical areas;
- *Resource allocation*: assist decision-makers in allocating funds and determining priorities;
- *Enforcement of standards*: determination of the extent to which legislative standards and existing criteria are being met or exceeded;
- *Trend analysis*: identification of overall trends in environmental quality over a period.

Over the years, many water quality indices have been developed (Ott 1978, for a complete treatise of environmental indices in general, and House and Ellis 1980 for an evaluation of indices used in the UK). The US National Science Foundation Water Quality Index (NSF WQI), which is one of the best-known and most commonly used general-purpose indices, is briefly discussed here. The NSF WQI was developed using a formal procedure to combine the opinions of a large panel of water experts throughout the USA. These experts were asked (i) initially to consider 35 water constituents for possible inclusion in the index, (ii) subsequently to review his/her original ratings after being informed of the other experts' rating and then (iii) to develop a rating curve for each of the 9 constituents considered to be of the greatest importance. Each of these curves bears a distinct shape depending on the affects the constituents cause at their different levels. The rating curves for DO and faecal coliforms are shown in Figs. 1a and 1b respectively (Ott 1978) as examples. For pesticides and toxic substances, if any substance exceeds its assigned upper limit the NSF WQI is set to zero. Based on the ratings by respondents, the weights of the 9 constituents were determined as shown in Table 1.

The index is calculated by determining the subindex value from the appropriate rating curve for each water constituent. The NSF

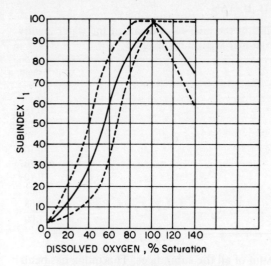

FIGURE 1a. Subindex function for DO
in the NSF WQ1

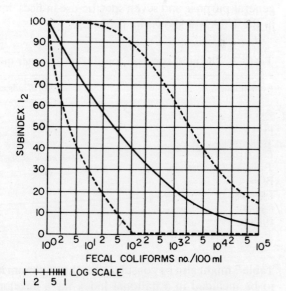

H—++HIIHH LOG SCALE
1 2 5 1

FIGURE 1b. Subindex function for faecal
coliforms (average number of organisms
per 100 ml) in the NSF WQ1

TABLE 1. Relative weights of the nine constituents of the water quality index developed by National Science Foundation

Constituent	Weight
DO	0.17
Faecal coliforms	0.15
pH	0.12
BOD$_5$	0.10
Nitrates	0.10
Phosphates	0.10
Temperature	0.10
Turbidity	0.08
Total solids	0.08
Total	1.00

WQI is the weighted sum of all the subindices. This index has been field-tested and has been widely used in various geographical areas in the USA, even though it has not received the status of a national index. The most common water quality constituents used in six general-purpose and seven specific-use indices developed and used in the USA are shown in Table 2.

TABLE 2. The most common constituents of water quality indices

Constituent	No. of indices using this constituent
DO	11
pH	11
Temperature	9
Faecal coliforms	8
Turbidity	7
BOD	6
Chlorides	6
Colour	6
Nitrates	6

Table 2 might also be considered as the minimum list of constituents to be included in a national index for a developing country. By examining the criteria proposed in the indices and by using mathematical analysis, Ott (1978) compiled a list of criteria for an ideal quality index. These criteria are the following:

1. logical scientific rationale or procedure behind index development;
2. reasonable balance between oversimplification and technical complexity;
3. sensitivity to small changes in water quality:
4. no eclipsing;
5. no ambiguity;
6. linearity in the aggression process;
7. dimensionless;
8. clearly defined range;
9. understanding of the significance of data;
10. ease in application;
11. ease in accommodation of new variable;
12. possible probability of interpretation;
13. variables that are widely and routinely measured;
14. inclusion of toxic materials;
15. inclusion of variables having clear effects on aquatic life or recreational use;
16. testings in a number of geographical areas;
17. reasonable agreement with expert opinion;
18. reasonable agreement with biological measures of water quality;
19. compatibility with water quality standards;
20. inclusion of guidance on how to handle missing values; and
21. clear documentation of limitations;

Biological Indices

In addition to indices which directly measure water quality, ecological indices are also sometimes used. These indices measure the response of key species of groups of organisms to pollution. There have been generally two approaches to the development of ecological indices.

(i) Pollution Indices. These are derived from observations of the response to pollution of affected species. These indices incorporate such factors as the relative abundance of indicator species, the reduction in community diversity, and the progressive loss of certain groups in response to pollution.

(ii) Diversity Indices. These are based on theoretical concepts of the structure of ecological communities. The degree of pollution is assessed by measuring the extent to which the community structure differs from that of an assumed model. In normal communities there are a few species represented by a large number of individuals, smaller numbers of several species, and many species which are represented only by very few individuals. Models have been established to describe this observation, and they are used as baseline data for pollution impact assessment. Any variation in terms of the number of species represented by a given number of individuals within a sample will indicate impact.

Hellawell (1978) has reviewed various ecological indices. As a general rule, ecological indices are highly site-specific and require skills of taxonomy for accurate identification and numeration. A clear knowledge of the local ecosystem is also necessary to establish base-line information for any 'before and after' study.

WATER QUALITY MODELLING

Water quality models are used to describe three principal phenomena: hydrological, thermal, and biochemical. Hydrological models essentially deal with the dynamics of the hydrological cycle. They are the most classical of the three types and play an important role in water development management.

Most of the biochemical water quality models in use today originated from two simple equations proposed by Streeter and Phelps in 1925 when they studied the pollution of the Ohio River. They observed that the BOD concentrations measured at successive points along the Ohio River below major discharges decreased as a function of distance travelled or time of flow. Accordingly BOD levels could be predicted by using a simple mathematical relationship based on time and a deoxygenation constant. The fundamental assumption in the Streeter–Phelps equations is that BOD and DO are sufficient to describe the biochemical processes of water pollution and natural purification. The model based primarily on this equation are helpful in predicting the BOD and DO concentrations or deficits as a result of a discharge of biodegradable wastes into a flowing water body. Solution of the Streeter–Phelps equation with different values of points in time produces a profile of DO known as the dissolved *oxygen sag curve* and is shown in Fig. 2. Beginning at a discharge point and moving downstream to the right, the DO decreases because deoxygenation occurs at a rate higher than reaeration. The deoxygenation rate increases in response to the increase

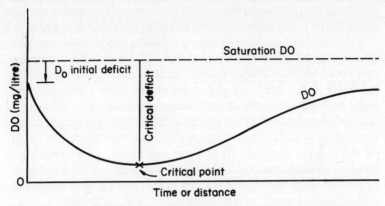

FIGURE 2. Example of an oxygen sag curve below a point-source discharge

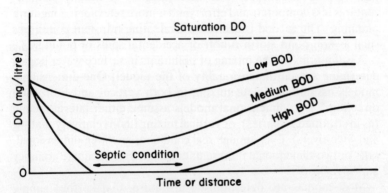

FIGURE 3. The effect of changing the BOD discharge into a stream on the oxygen sag curve

oxygen deficit. When the two rates become equal, the DO level reaches its lowest level—the critical point. From this point on, the re-aeration rate exceeds the deoxygenation rate, and the DO gradually recovers to the initial level. The exact shape of the oxygen sag curve will depend on the initial load of BOD at the discharge point, as shown in Fig. 3. Obviously there should be a careful balance between benefits and costs of wastewater treatment for BOD reduction. Too high a BOD load will reduce the DO concentration to zero, which will give rise to septic conditions, while too great of BOD reduction is too costly. Medium BOD loads with moderate wastewater treatment will save costs while natural assimilation can be relied upon to purify the rest of the BOD discharged.

Some models based on the Streeter–Phelps concept also consider some additional biochemical phenomena. For example, the assumption in the Streeter–Phelps model that the deoxygenation rate and the BOD decay rate are equal is not always true. The decay rate could be higher than the deoxygenation rate because of sedimentation, or lower because of resuspension. Thus, first-order exponential decay dilution and sedimentation models have been developed based on the concept proposed by Streeter and Phelps. More and more complex multi-parameter water quality models can be developed to predict more accurately the physical, chemical and biological interactions of many aquatic constituents and organisms.

Water quality models can be used for steady-state or dynamic time-varying conditions. The first category is usually simpler, requires less computational effort, and is more relevant to long-term planning. The second is useful for evaluating transient conditions such as non-point storm runoff or accidental spills of pollutants.

Assumptions on the mixing of pollutants in surface water bodies determine spatial dimensionality of the model. One-dimensional models assume complete mixing in both vertical and horizontal directions. Two-dimensional models assume either lateral mixing (as in stratified estuaries), or vertical mixing (as in relatively shallow and wide-rivers). Even though real conditions are three-dimensional, one- or two-dimentional representations may give adequate accuracy and save costs.

Most models are deterministic in nature, that is, they simply give estimates of mean values of the constituents. Stochastic or probabilistic models take into account the probability distributions— not just the expected values—of various processes occurring in nature. These models require elaborate data inputs which are not always available. Model solution methodology varies from manual computation to the use of nomographs and computers. Most computer models are simulation type which provide the values of water quality parameters for given flows, pollutant discharge rates and the degree of pollution control. For each set of assumptions, a trial-and-error simulation run is performed until satisfactory results are obtained. The models are solved either by integrating basic differential equations, or by using numerical analysis techniques such as finite difference or finite element.

If cost or time constraints dictate against simulation models, optimization models may be used. The most common solving methods are linear programming and dynamic programming, but non-linear or mixed-integer programming is sometimes used. Optimization models are more limited in scope than simulation models, but they can help to define the range of acceptable alternatives that best satisfy pre-determined objectives. These alternatives may be double-checked using a verified simulation model which provides a more accurate basis for economic and technological comparisons, Biswas (1976 and 1981) and Rinaldi *et al.* (1979) provide further details on water quality modelling. The current status of water quality modelling for various constituents is summarized in Table 3 (Biswas 1981).

TABLE 3. Current status of water quality modelling

Water quality constituents	Modelling status
Transport	
steady state	good
dynamic	marginal
BOD-DO	good
Total dissolved solids	moderate
Suspended solids	marginal
Bacteria	marginal
Synthetic chemicals	poor
Simple industrial chemicals	marginal
Simple metals	marginal
Complex metals	poor
Zooplankton	poor
Viruses	poor
Turbidity	poor
Colour	poor
Algae	marginal
Nutrients	moderate
Temperature	good
Synergistic models	impossible

LAKE WATER QUALITY

Water quality management for lakes (including man-made reservoirs) is different from that for streams. Only some basic issues will be discussed here, but more detail can be seen in Jørgensen (1980).

Water inputs to lakes include streams, surface runoffs, groundwater inflows, and atmospheric precipitation. Chemically, consideration of the water quality of a lake is somewhat similar to that of the streams feeding the lake, but additional factors profoundly modify the water quality characteristics of a lake. Reduced velocity and turbulence allows a substantial portion of the solid particles brought in streams to settle, reducing the turbidity of the lake water. Also, internal mixing and the longer retention time reduce the range of quality variations far below those observed in streams. The result is a very stable aquatic ecosystem in a lake, which can be easily upset by any sudden change in the water quality regime. For instance, reduced turbidity results in increased light penetration, which could promote algal growth, which in turn changes the ecological balance in a way that is entirely different from that of streams.

A phenomenon that causes great impact on the quality of lake water as well as of the water downstream is thermal stratification. In deep lakes (more than 30 m) the top layer (*epilimnion*) becomes warmer than the bottom layer (*hypolimnion*) during the summer. For this reason, only water in the top layer circulates, and it does not mix with water in the colder bottom layer. A steep temperature gradient is created between the two layers, and the transitional zone is called the *thermocline*. If the thermocline is below the depth of light penetration, no oxygen production from photosynthesis will occur, resulting in constant anaerobic conditions in the bottom layer. This condition is called *summer stagnation*.

In temperate climates, during the autumn, the temperature of the upper layer drops until it is the same as of the bottom layer, resulting in circulation of the entire lake, and oxygen is brought from the surface to the bottom. When the upper layer temperature drops to 4 °C, it becomes heaviest and so sinks to the bottom, creating a *fall overturn*. Then, as spring arrives, the upper layer temperature increases from the freezing level up to 4 °C, which makes it sinks to the bottom, creating a *spring overturn*. Figures 5 and 6 depict the seasonal profiles of temperature and oxygen respectively.

For tropical climates, the situation is quite different. In areas where the temperature never falls to 4 °C there is no fall or spring overturn. In comparison to temperate climates, the summer conditions exist year-round. In other words, the *summer stagnation* is permanent: only condition (2) exists in Figs. 4 and 5. The deeper the lake, the

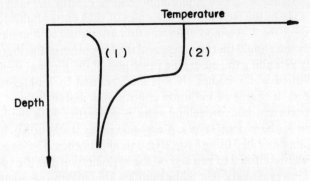

FIGURE 4. Temperature profile in a stratified lake:
(1) winter condition; (2) summer condition

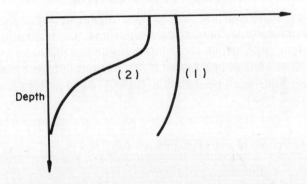

FIGURE 5. Oxygen profile in a stratified lake:
(1) winter condition; (2) summer condition

worse is the stagnation. Thus, without proper management deep man-made reservoirs usually have very large volumes in which the water quality is deteriorated and fishery is not productive

Mathematical modelling of lakes and reservoirs is a difficult process. Response times in lakes are long, many variables cannot be controlled or measured, and natural factors such as weather complicate the natural and biological processes in the water environment. Despite these constraints, under certain conditions modelling of lakes can prove to be useful for water quality management.

Modelling of the lake water quality in the 1960s was initially

206 / N. C. THANH, D. M. TAM

meant to study severe eutrophication of some natural and man-made lakes. One of the well-known examples of such early application is the Lake Erie. Lake modelling was first attempted by limnologists to study the annual thermal cycle and the formation of the thermocline, by extending the modelling principles of the thermal structure of stratified water bodies. It was later refined by biologists and engineers in charge of pollution control and abatement.

Most current lake modelling work is concerned with the finite-element for horizontal layers. Mass and energy transports between layers are described using hydrodynamic approaches. These models are usually confined to two layers, the hypolimnion and the epilimnion. However, very few real examples are currently available of mathematical models successfully applied to lake water quality management.

WATER POLLUTION CONTROL

Various options available for collection, treatment and disposal of wastewater for water pollution control will be briefly discussed in this section. More details on mainly conventional technologies can be found in Metcalf and Eddy (1979). Various options for water pollution control are presented in Table 4 (adapted from Kneese and Bower 1968).

TABLE 4. Options of water pollution control

Principle	Method	Example
A. Reducing waste generation	1. In-plant recirculation of water	Water used for cooling purposes is collected for recirculation and reuse
	2. Segregation of concentrated waste streams	Removal of solid wastes by hand or mechanical means, to be disposed of easily in a separate way, rather than flushing them down the gutter which makes the wastewater stream more difficult to treat
	3. Waste elimination	Use residues of agro-industries (e.g. pineapple) to feed animals (e.g. pigs)
	4. Change type of raw material inputs	May be difficult
	5. Change production process	Change from pressure to solvent extraction to achieve better oil extraction efficiency in vegetative oil industry, reducing oil load in waste stream

Principle	Method	Example
	6. Change or modify product outputs	Detergents with lower contents of phosphorus. Change from disposable to recyclable packing material in packaging and bottling industries
B. Reducing wastes after generation	7. Material recovery	Collection of palm fruit fibres in palm oil industry, to be dried and used as fuel for boilers
	8. Byproduct production	Integrated farming, which consists of duck/pig production, aquaculture and crop production
	9. Waste treatment	Oxidation ditch, aerated lagoon, waste stabilization pond, land treatment, wetland
	10. Effluent reuse	Aquaculture, composting, irrigation and fertilization, biogas production
C. Improve assimilative capacity of receiving water	11. Addition of dilution water	
	12. Multiple outlets from reservoirs	
	13. Reservoir mixing	
	14. Reaeration of streams	
	15. Saltwater barriers	
	16. Effluent redistribution	
D. Corrective measures on ecosytem	17. Chemical treatment in reservoirs	Application of chemicals (e.g. aluminum sulphate) to precipitate phosphorus
	18. Reduction of biomass concentration	Introduction of herbivorous animals to graze aquatic weeds, or spraying of herbicides to control algae and weeds

For developing countries, the principles of reducing waste generation and improving wastes recovery are the most appropriate and feasible processes. In this regard the situation in many developing countries could be substantially improved. In almost every pollution survey, it has been observed that more pollutants are discharged than necessary, and with only minimal costs carrying out methods

1–3 of principle A, the pollutant load could be significantly reduced. The cost–benefit ratio in this case would be very favourable. And with more investment for methods 5–6, the situation would be changed even more dramatically. For example, in one palm oil plant in Thailand it was estimated that the amount of oil discharged in the effluent was as much as one-fourth of the plant's oil yield. Local people collected the effluent and boiled it over a open fire to recover the oil contents. With some modifications in the manufacturing process to increase the efficiency of oil extraction, the plant could have obtained more yields. Such modifications are not only cost-effective but also substantially contribute to the preservation of environmental quality.

Wastewater Treatment

There are a variety of wastewater treatment processes, but for developing countries, one of the most appropriate is a waste stabilization pond system (Biswas and Arar 1988; WHO EMRO 1987). The main driving force in this process is the sun, and natural organisms consisting of algae, bacteria and plankton help break down the organic matter and remove pathogens in the wastewater. A pond system typically consists of a series of 3–5 ponds. The first pond may be either anaerobic or facultative depending on the organic loading rate. Anaerobic ponds are more efficient in terms of pollutant mass removed per unit of land area, but could be objectionable to a nearby community. Most organic removal takes place in the first one or two ponds, whereas pathogen die-off in the remaining ponds, reducing the pathogen levels in the effluent to an acceptable level. The performance of stabilization pond systems is superior to many conventional treatment processes: organic removal efficiency can reach 90 per cent or more, while pathogens are removed by around 10,000 times. The only drawback in terms of effluent quality is the high levels of suspended solids which consist mainly of live algal cells, thus violating common effluent standards. High levels of algae may deteriorate the quality of downstream water.

Wastewater Reuse

Aquaculture using wastes has been practiced traditionally in many regions of the world (Edwards 1980; 1985). Throughout Asia, fish and aquatic plants have been grown in waters contaminated with faecal matter. Fishponds near Calcutta receive sewage from the city, and the harvested fish are sold back to the city. Latrines are built over fish ponds in China and Vietnam, fish raised in cages submerged in faecal-contaminated streams in Indonesia, and vegetables (such as the water spinach) grown in canals fed with sewage in Thailand. Some public health threats may exist in such practices, especially when appropriate precautions are not taken. However, some important points should be kept in mind when dealing with this issue. First, these practices help produce substantial quantities of low-cost protein food for the poor local people. Any option that wipes out this protein source without adequate replacement will be disastrous. Second, for some systems such as overhung latrine-fish pond, the pollutants are contained quite well. An alternative waste management option which partially removes the pollutant load but advertently spreads the remainder will not necessarily be better. A more rational strategy is to use waste to grow aquatic plants, and use the plants to grow human food. Aquatic plants can be used in many other ways, and the various options are reviewed by Polprasert *et al.* (1986).

The cultivation of the water fern *Azolla* sp. as a green manure deserves attention. This aquatic plant grows in symbiosis with the blue-green algae Anabaena azollae, which is able to form its protein by fixing nitrogen from the air. The fern is grown in rice fields before the rice-growing season, and is fertilized with animal wastes. Due to its fast growth (doubling time of 3–10 days), a high amount of nitrogen fertilizer can be produced quickly, to be used later by rice paddy. Azolla cultivation and use techniques have been developed in China and Vietnam, and have received attention elsewhere for applied research and development (Polprasert *et al.* 1986).

Algal culture in shallow, high-rate ponds fed with treated wastewater has received a fair share of research, but its economic viability is still not favourable. The main problem lies with difficulty in parasite and predator control, the unavailability of affordable techniques for separating tiny algal cells from water and for processing (drying) the algae for marketing, and marketing itself. In some areas in China, the green alga *Chlorella* and the blue-green *Spirulina* have been cultivated with animal wastes, and the algal slurry is fed directly to pigs. The state-of-the-art of algal biomass production and use can be found in Shelef and Soeder (1980).

Irrigation/fertilization also includes land application as a treatment process. Throughout Korea, China, Taiwan, Japan, Vietnam and to a certain extent, Thailand, the Philippines and Malaysia, nightsoil has been applied directly onto agricultural land. Sewage irrigation is also applied in several developing countries. In India, Bombay initiated sewage farming in 1877, followed by Delhi in 1913. The daily volume of sewage reaches 3.6 million m³, of which about 50 per cent is being utilized for crop irrigation. Near Mexico City, about 40 m³/s of sewage is used to irrigate 41,500 ha of fodder and grain not for human consumption. Some 16,000 ha grown with vegetables and salad crops receives the raw wastewater from the city of Santiago, Chile. These reuse practices have been done more by economic incentive than by environmental awareness. As a result, the environment does not necessarily benefit from it.

Wastewaters often contain higher levels of metals than natural waters do. For this reason, metals should be closely monitored to determine the use and management practices required. The impact of metals in irrigation water on soil and biota is summarized in Table 5.

With regard to public health threats posed by wastewater irrigation, an extensive investigation of available data (Biswas and Arar 1988;

TABLE 5. The impact of metals in irrigation water on soil and biota

Metal	Recommended maximum concentration	Impact on soils and biota
Aluminium (Al)	5.0	Can cause non-productivity in acid soils (pH<5.5), but more alkaline soils at pH>7.0 will precipitate the ion and eliminate any toxicity
Beryllium (Be)	0.10	Toxicity to plants varies widely, ranging from 5 mg/L for kale to 0.5 mg/L for bush beans
Cadmium (Ca)	0.01	Toxic to beans, beets and turnips at concentrations as low as 0.1 mg/L in nutrient solutions. Conservative limits recommended due to its potential for accumulation in plants and soils to concentrations that may be harmful to humans.
Cobalt (Co)	0.05	Toxic to tomato plants at 0.1 mg/L in nutrient solution. Tends to be inactivated by neutral and alkaline soils
Chromium (Cr)	0.20	Not generally recognized as an essential growth element. Conservative limits recommended due to lack of knowledge on its toxicity to plants
Copper (Cu)	0.20	Toxic to a number of plants at 0.1 to 1.0 mg/L in nutrient solutions

Metal	Recommended maximum concentration	Impact on soils and biota
Iron (Fe)	5.0	Not toxic to plants in aerated soils, but can contribute to soil acidification and loss of availability of essential phosphorus and molybdenum. Overhead sprinkling may result in unsightly deposits on plants, equipments and buildings.
Lithium (Li)	2.5	Tolerated by most crops up to 5 mg/L; mobile in soil. Toxic to citrus at low concentrations (<0.075 mg/L). Acts similarly to boron
Manganese (Mn)	0.20	Toxic to a number of crops at a few tenths to a few mg/L, but usually only in acid soils
Molybdenum (Mo)	0.01	Not toxic to plants at normal concentrations in soil and water. Can be toxic to livestock if forage is grown in soils with high concentrations of available molybdenum.
Nickel (Ni)	0.20	Toxic to a number of plants 0.5 to 1.0 mg/L; reduced toxicity at neutral or alkaline pH
Lead (Pb)	5.0	Can inhibit plant cell growth at very high concentrations.
Selenium (Se)	0.02	Toxic to plants at concentrations as low as 0.025 mg/L and toxic to livestock if forage is grown in soils with relatively high levels of added selenium. An essential element to animals but in very low concentrations.
Vanadium (V)	0.10	Toxic to many plants at relatively low concentrations
Zinc (Zn)	2.0	Toxic to many plants at widely varying concentrations; reduced toxicity at pH>6.0 and in fine-textured or organic soils

Shuval *et al*. 1986) has indicated that previous public health positions have been conservative. The main public health concern in wastewater irrigation in developing countries would be helminths, not bacteria or viruses. This problem could be alleviated by holding the wastewater in a settling tank/pond for 2 days before using the effluent for irrigation. A stabilization pond system with a minimum 20-day retention time can remove almost all bacteria and viruses and can produce an effluent suitable for unrestricted irrigation of vegetables. Wastewater irrigation offers various benefits of increased agricultural/forestry yields, improved food supply, reduced environmental pollution and spin-offs such as job creation and rural development. In developing countries, these benefits have not been exploited to any extent while wastewaters are allowed to pollute the local water bodies.

There are three basic land application methods, namely slow rate, overland flow, and infiltration-percolation. These methods are depicted in Fig. 6. The slow-rate process is easier to apply in most conditions, since overland flow requires a gentle slope over the application area and infiltration-percolation is suitable only where the infiltration rate is high such as in sandy soils. Slow-rate irrigation is based on the idea of utilizing both the water and the nutrients of wastewater for crop production. More details on wastewater reuse in irrigation/fertilization can be found in Shuval *et al.* (1986) and Biswas and Arar (1988).

WATER QUALITY MANAGEMENT FROM
NATIONAL PERSPECTIVES

Main issues at the national level for water quality management are summarized as follows (United Nations 1987):

- Water quality management legislation should be consistent with national requirements of economic and social realities, and human resource capabilities.
- Water quality legislation should be enforceable. An adequately staffed, equipped and functional monitoring system is crucial. The political will to prosecute violators should be created. Appropriate effluents standards should be adopted, and discharge levels of certain substance into water courses should be restricted. Permits should be issued only to economic activities using 'clean' processes or located in areas with adequate environmental assimilative capacity.
- Water quality management within the context of legislative actions may include:
    creation of an appropriate institutional infrastructure;
    requirements for industry and municipalities to treat their effluents in order to ensure water quality standards in surface and groundwater are not violated;
    designation of protected areas or hydrological regions and aquifers;
    requirements for adopting special designs or technology to

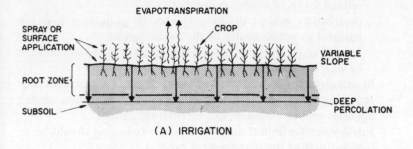

(A) IRRIGATION

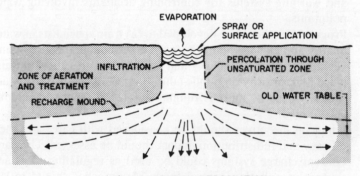

(B) INFILTRATION PERCOLATION

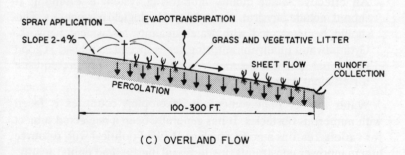

(C) OVERLAND FLOW

FIGURE 6. Methods of land application

reduce levels of wastes generated;

prohibition against discharging specific pollutants that are harmful to public health or the environment;

control of production, processing, transportation and storage of water pollutants;

- Mechanisms for coordination of water quality management programmes to ensure reduction of duplication of efforts among agencies and to eliminate institutional infighting.
- Environmental impact assessment and monitoring should be an integral part of the management process.
- Availability of an efficient and well-equipped emergency services and warning systems for controlling accidents involving water pollutants.
- Programmes should be developed to (a) train specialists in water quality and management; (b) teach environmental issues in universities, technical colleges and schools, and (c) focus on and place high priority on public education. Mass media techniques should be used, and environmental interest groups should be encouraged.
- Financial incentives should be considered with care, and their efficacy and redistribution impact should be assessed. User and effluent charge systems could be used as regulating tools and sources of pollution control funds. Approaches should reflect the specific local conditions and charges should be updated when appropriate.
- An effective water quality monitoring system is essential. It should include physical, chemical and biological parameters. This should be integrated with water quantity assessment network. Groundwater quality monitoring needs special emphasis. A good functional data management system is a necessary prerequisite for efficient water quality management.

Water quality management in developing countries is faced with numerous obstacles. It has generally been a neglected subject for various reasons among which are lack of political will, resource and manpower constraints, institutional inertia, and public apathy. As a crucial first step, public and political awareness should be raised so that it receives more attention, since water quality directly influences many issues; human health, agriculture, aquaculture, industry, and tourism—all of which are vital to national development.

Decision-makers should be convinced that investments in enhancing the quality of the nation's water will directly contribute to social and economic benefits, just like investments in other sectors. For example, in the Caribbean region, the lack of a comprehensive wastewater management or improper waste disposal has led to wide-scale pollution of tourist resorts, which has slowed down or even frozen further development. This could happen in other countries.

Determination of water quality objectives and criteria is the next important step. In many cases, standards are directly adopted from the West or from those recommended by the WHO, without considering the country's social, economic, cultural or climatic requirements and the manpower expertise and institutions necessary to implement them. The agency or the polluter cares much about enforcement and compliance since it is clear that those unrealistic standards cannot be achieved anyway.

This is perhaps rooted in the mentality that lower standards appropriate to the local conditions are insulting 'second rate', while it is desirable to have the high standards just like in any industrialized country. Developing countries should consider realistically what can be achieved and then take appropriate actions to enforce them. Otherwise best will continue to be the enemy of good.

The feeling against something 'second rate' also exists with innovative, low-cost appropriate technology. Despite much effort during the past two decades to promote waste stabilization ponds, it is still often difficult to convince government authorities that this technology is as good as, if not better than, conventional treatment processes, especially for countries in tropical and semi-tropical regions.

Blind adoption of western technology for water quality management has created more problems than they have solved. Technology implementation requires supporting 'software' and management. Among these are institutional strengthening, resources and expertise available, public information and motivation, community participation, including involvement of women. For instance, many industries still release unnecessarily high pollution loads into the environment. With better information motivation and minimum investment, levels of pollution could be substantially reduced.

This leads back to an old golden rule: prevention is better than treatment. At many factories storm runoff is mixed with process wastewater, and this makes treatment so costly that the factory cannot afford and the authorities cannot enforce it. Large amounts of coarse solid particles are swept into the wastewater streams while they can be collected separately and easily disposed of by low-cost landfilling. To correct this situation, awareness, public information and motivation are necessary.

By combining awareness, realistic planning and decision-making, software supporting activities, and waste control at source, developing countries can have a good start, and many of their water quality problems would be greatly alleviated within a reasonable time frame.

REFERENCES AND BIBLIOGRAPHY

Adam JWH. 1980. 'Health aspects of nitrate in drinking water and possible means of denitrification (a literature review)', *Water SA*, vol. 6, no. 2, pp. 79–84.

*Alabaster JS (ed). 1980. *Water quality criteria for freshwater fish*, Butterworths, London.

Arthur JP. 1983. 'Notes on the design and operation of waste stabilization ponds in warm climates of developing countries', World Bank Technical Paper no. 6, The World Bank, Washington DC.

*Biswas AK (ed). 1976. *System approach to water management*, McGraw-Hill, New York, NY.

*—— (ed). 1981. *Models for water quality management*, McGraw-Hill, New York, NY.

Biswas AK, Arar A (eds). 1988. *Treatment and reuse of wastewater for irrigation*, Butterworths, Guilford, UK.

Bitton G, Farrah SR, Montague CL, Akin EW. 1986. 'Viruses in drinking waters', *Environmental Science and Technology*, vol. 20, no. 3, pp. 216–22.

Cairns J, Patrick R (eds). 1986. *Managing water resources*, Praeger, New York, NY.

Compbell IC. 1981. 'A critique of assimilative capacity', *Journal of Water Pollution Control Federation*, vol. 5, no. 5, pp. 604–7.

*Canter LW. 1985. *River water quality monitoring*, Lewis Publishers, Michigan, USA.

Davis GH. 1987. 'Effect of energy development on water quality', *Nature and Resources*, vol. 23, no. 1, pp. 2–17.

De Koning HW. 1981. 'Rapid assessment of air, water and land pollution sources', *Environmental Monitoring and Assessment*, vol. 1, no. 2, pp. 129–41.

Drinking Water Research Division, USEPA. 1981. 'Chlorine, is there a

Documents marked with an * are key reading materials.

better alternative?', *The Science of the Total Environment*, vol. 18, pp. 235–43.

Economic and Social Council. 1985. *Water resources: Progress achieved and prospects in the implementation by Governments of the Mar del Plata Action Plan—Report of the Secretary General*, E/C.7/1985/5, United Nations, New York.

Economic Commission for Western Asia. 1982. *Development of guidelines for efficient water management in the ECWA Region*, Economic Commission of Western Asia, Baghdad, Iraq.

Edwards P. 1980. 'A review of recycling organic wastes into fish, with emphasis on the topics', *Aquaculture*, vol. 21, pp. 261–79.

———. 1985. 'Aquaculture: A component of low-cost sanitation technology', Technical Paper no. 36, The World Bank, Washington DC.

Ewusie JY. 1980. *Elements of tropical ecology*, Heinemann Educational Books, London, UK.

Geldreich EE. 1976. 'Fecal coliform and fecal streptococcus density relationships in waste discharges and receiving waters', *CRC Critical Reviews in Environmental Control*, October, pp. 349–69.

*Hellawell JM. 1978. *Biological surveillance of rivers: A biological monitoring handbook*, Water Research Centre, Stevenage, UK.

House M, Ellis JB. 1980. 'Water quality indices: An additional management tool', *Water Science and Technology*, vol. 13, no. 7, pp. 413–23.

*IDH-WHO Working Group on the Quality of Water. 1978. 'Water quality survey', *Studies and Reports in Hydrology*, no. 23, United Nations Educational, Scientific and Cultural Organisation, Paris, France and World Health Organisation, Geneva, Switzerland.

Jørgensen SE. 1980. *Lake management*, Pergamon, New York, NY.

Kneese AV, Bower BT. 1968. *Managing water quality: Economics, technology, institutions*, The Johns Hopkins Press, Baltimore, USA.

*Lamb JC. 1985. *Water quality and its control*, John Wiley, New York, NY.

Lin SD. 1977. 'Tastes and odors in water supplies—A review', Circular 127, Illinois State Water Survey, USA.

Lund HF (ed). 1971. *Industrial pollution control handbook*, McGraw-Hill, New York, NY.

*McJunkin FE, 1982. 'Water and human health', US Agency for International Development, Washington DC.

Metcalf and Eddy. 1979. *Wastewater engineering—Treatment, disposal, reuse*, McGraw Hill, New York, NY.

Miller KJ. 1985. 'Water reuse practices in the United States' in *Australian Water and Wastewater Association 1985 International Convention Proceedings*, April 28–May 3, Melbourne, Australian Water and Wastewater Association, Melbourne, pp. 101–14.

National Academy of Sciences. 1977. *Drinking water and health; Volumes 1–5*, Safe Drinking Water Committee, National Academy of Sciences, Washington DC.

218 / N. C. THANH, D. M. TAM

*Ott WR. 1978. *Environmental indices—Theory and practice*, Ann Arbor
Science Publishers, Michigan, USA.
Pascoe D, Edwards RW. 1984. 'Freshwater biological monitoring',
Proceedings of a Specialised Conference, Cardiff, UK, 12–14 September,
Pergamon, New York, N.Y.
Pescod MB. 1977.'Surface water quality criteria for tropical developing
countries' in *Water, wastes and health in hot climate*, John Wiley, New York,
NY.
Phillips VR. 1986. 'Remedies to problems caused by agriculture—The
engineering solution' in *Effects of land use on fresh waters: Agriculture,
forestry, mineral exploitation urbanisation* (ed. JF de LG Solbe), Ells
Horwood Ltd., Chichester, UK.
Polprasert, C, Sikka, B, Tchobanoglous, G. 1986. 'Aquatic weeds and
their uses—An overview of perspectives for developing countries',
Environmental Sanitation Review, no. 21.
*Rinaldi S, Soncini-Sessa R, Stehfest H, Tamura, H. 1979. *Modelling and
control of river quality*, McGraw-Hill, New York, NY.
*Schaller FW, Bailey GW (eds). 1983. *Agricultural management and water
quality*, Iowa University Press, Ames, Iowa, USA.
Schmidtke NW, Salloum I. 1985. 'Nutrient control technology in Canada'
in *Australian Water and Wastewater Association 1985 International
Convention Proceedings*, April 28–May 3, Melbourne, Australian Water
and Wastewater Association, Melbourne, pp. 385–95.
Shelef G, Soeder CJ (eds). 1980. *Algae Biomass—Production and use*.
Elsevier/North-Holland Biomedical Press, Amsterdam.
Shuval HI, Adin A, Fattal B, Rawitz E, Yekutiel P. 1986. *Wastewater
irrigation in developing countries—Health effects and technical solutions*,
World Bank Technical Paper no. 51, The World Bank, Washington DC.
Somers E. 1981. 'Environmental monitoring and the development of
health standards', *Environmental Monitoring and Assessment*, vol. 1
no. 1, pp. 7–19.
Tam DM, McGarry MG, Brocklehurst C. 1986. 'Recent scientific and
technical developments in the Canadian water supply and their possible
relevance to developing countries', Unpublished report, Cowater International, Ottawa, Canada.
Truesdale G, Taylor G. 1978. 'Quality implications in reservoirs filled
from surface water sources', *Progress in Water Technology*, vol. 10,
nos. (3/4), pp. 289–300.
United Nations. 1987. *Final Report—Interregional symposium on improved
efficiency in the management of water resources: Follow-up to Mar del
Plata Action Plan*, United Nations, New York, NY.
Wheater DWF, Mara DD, Oraguli JI. 1979. 'Indicator systems to distinguish
sewage from stormwater run-off and human from animal faecal material'
in *Biological indicators of water quality*, John Wiley, Chichester, UK.
Whipple W. 1977. *Planning of water quality systems*, Lexington Books,
Massachusetts, USA.

WHO. 1984. *Guidelines for drinking-water quality', Vol. 1: Recommendations; Vol. 2: Health criteria and other supporting information*, World Health Organisation, Geneva, Switzerland.

WHO EMRO. 1987. *Wastewater stabilization ponds—Principles of Planning and practice*, WHO EMRO Technical Publication no. 10. World Health Organisation Regional Office for the Eastern Mediterranean, Alexandria, Egypt.

Groundwater Quality Management

R. MACKAY

INTRODUCTION

In most parts of the world, groundwater has generally been considered to be a readily available, good quality source of water for drinking and for agricultural and industrial uses. However, with increasing demands, significant changes in land use patterns and a vast increase in the quantities and types of industrial, agricultural and domestic effluents entering the hydrological cycle, the stresses on groundwater, both in terms of quantity and quality, are growing rapidly. In the past, groundwater managers have primarily been responsible for establishing the 'safe yield' of an aquifer system with little interest in how groundwater quality might change with time as a result of human activities. In reality, determination of the 'safe yield' for a groundwater resource must consider both its quantity and quality. Thus, over the last two decades there has been a substantial move towards a more complete understanding of groundwater contamination and how the impacts of contamination can be mitigated by appropriate management actions and technological developments.

In order to be able to develop an appropriate management plan for an aquifer system which takes into account the quality of the resource, the factors influencing the control of groundwater quality must be understood. Only with this knowledge can the groundwater manager assemble and interpret all the relevant data before deciding on a particular strategy. Assembling the data base depends largely on correctly formulating the problem. This may be undertaken by developing a series of questions appropriate to the definition of the problem. For example, an obvious question should be—'When does contamination become pollution?' Contamination is used here to indicate the presence of chemical, bacterial or radiological species in the groundwater at levels greater than the background values, whilst pollution indicates the presence of species at levels harmful in terms of the use of the water. To answer this question properly, a range of further questions need to be posed and answered. Among these questions are:

- What contaminations are present in the groundwater?
- What are harmful levels for specific contaminants?
- How will the contaminant levels in the groundwater system change with time?
- Can the contamination levels be controlled, e.g. by altering abstractions?
- Can the source of contamination be controlled effectively?
- How will future plans for development and use of the aquifer system be affected by the contamination?

In the following sections an attempt is made to establish a basic framework for providing answers to the above questions and to define strategies for the management of groundwater quality.

Groundwater quality management is a relatively new field and as such there is still much research to be undertaken before an 'optimal' approach to groundwater management can be defined. However, given present stresses on groundwater resources in most countries of the world it is important that resource managers begin to come to terms with the very real problems of quality management. Managers must recognize the complex, multi-disciplinary nature of the problem and the need to enlist the skills of experienced specialists in the fields of engineering, hydrogeology, biology, chemistry and agriculture in order that approprite solutions to groundwater pollution can be established.

Consideration is thus given in this chapter to the potential sources of groundwater pollution, the physical characteristics of groundwater systems controlling contaminant accumulation and migration, and the methods of monitoring and mitigating contamination. Possible approaches for regulating the potential for contamination are also presented.

SOURCES OF CONTAMINATION

Contamination may occur wherever the potential exists for transport of matter into the groundwater environment, either from the ground surface or by the drawing in of natural contaminations from adjacent subsurface or surface waters through human action. Whether the contaminant is considered as a pollutant will depend on the extent to which the use of the groundwater is affected by the presence of the contaminant. Drinking water quality guidelines (WHO 1987) define the perceived acceptable limits with regard to the physical, chemical, bacteriological and radiological content of potable water. Thus any contaminant present in the groundwater system which increases the need for purification or treatment may be considered to be a pollutant. Obviously, the standards set for industrial and agricultural water supplies are different from those for potable water, although the parameters adopted will be similar. Therefore, end use must be considered, both for present and future demands, to define acceptable contamination levels of a resource. There are a wide range of potential sources of contamination for groundwater. Table 1 modified from Todd *et al.* (1976), subdivides contamination sources into six categories. Against each category the potential hazard in terms of the primary parameters for water quality standards is presented. Tabulation of this type provides a useful indicator of the relative pollution potential of a contamination source against a particular water use. However, considerably greater detail is required to establish the actual potential for damage to the groundwater system.

A further classification of contamination sources arises from the spread of the source. Sources such as drains, landfills, septic tanks and oil storage tanks are considered to be point sources whereas applications of fertilizers and pesticides and saline groundwater bodies are diffuse or non-point sources. Whilst point sources potentially offer a greater hazard arising from higher pollutant concentrations, their effect is generally limited to a more localized domain within the groundwater system. Thus, greater scope often exists for management of an aquifer system to minimize the impacts of point sources compared with diffuse sources, where such sources are unavoidable.

Although avoidance of contamination from all the sources indentified in Table 1 would be the most satisfactory solution, such an option is neither practicable nor economically justified. This is particularly true where such sources are already in place. However, future planning must address the problem of minimizing the creation of such sources by improved industrial technology for waste re-use (Heath and Lehr 1987), better agricultural management (Schepers 1982) and better engineered containment of contaminants, including appropriate siting of waste facilities away from susceptible groundwater sources (LeGrand 1983). Only the last of these issues is addressed in the discussions that follow.

TABLE 1. Major sources of groundwater contamination and types of contaminants and their relative significance

Source	Type of pollutant				
	Physical	Inorganic chemical	Trace elements	Organic chemical	Bacterio-logical
Municipal					
Sewer leakage	Minor	Primary	Secondary	Primary	Primary
Sewage effluent	Minor	Primary	Secondary	Primary	Primary
Sewage sludge	Minor	Primary	Primary	Primary	Primary
Urban runoff	Minor	Secondary	Variable	Primary	Minor
Waste disposal	Minor	Primary	Primary	Primary	Secondary
Septic tanks and cesspools	Minor	Primary	Minor	Secondary	Primary
Agricultural					
Leached salts	Minor	Primary	Minor	Minor	Minor
Fertilizers	Minor	Primary	Secondary	Secondary	Minor
Pesticides	Minor	Minor	Minor	Primary	Minor
Animal waste	Minor	Primary	Minor	Secondary	Primary
Industrial					
Cooling water	Primary	Minor	Primary	Minor	Minor
Process waters	Variable	Primary	Primary	Variable	Minor
Water treatment plant effluent	Minor	Primary	Secondary	Minor	Minor
Hydrocarbons	Secondary	Secondary	Secondary	Primary	Minor
Tank and pipe-line leakage	Variable	Variable	Variable	Variable	Minor
Oilfield wastes					
Brines	Primary	Primary	Primary	Minor	Minor
Hydrocarbons	Secondary	Secondary	Secondary	Primary	Minor
Mining	Minor	Primary	Variable	Minor	Minor
Miscellaneous					
Surface water	Variable	Variable	Variable	Variable	Variable
Sea water intrusion	Primary	Primary	Primary	Minor	Minor
Transport	Minor	Minor	Minor	Primary	Variable

(Based on data from Todd *et al.* 1976)

For existing sources of contamination the real problems to be addressed are how to evaluate the future impact of the contamination on the groundwater environment and whether there is a need for action to arrest the problem and/or to rehabilitate the groundwater system.

GROUNDWATER FLOW AND CONTAMINANT TRANSPORT

Groundwater is an integral part of the hydrological cycle. Water enters a groundwater system by infiltration of surface waters through the ground surface to the saturated zone of the aquifer. Unless the top of the saturated zone is at or very close to the groundwater surface, the infiltrating water passes through an unsaturated zone where the pores or voids of the soil or rock mass contain air plus water vapour in addition to water in its liquid state. Under most circumstances water is considered to migrate vertically in the unsaturated zone. Once in the saturated zone the water moves laterally as well as vertically in response to hydraulic forces acting upon it until it leaves through the ground surface again or is removed through abstraction points, such as boreholes, or is released directly into the sea. Groundwater is defined to be only that part of subsurface water found in the saturated zone at pressures greater than atmospheric. Water in the unsaturated zone is frequently referred to as soil water (Heath 1983).

Groundwater movement is dependent on the hydraulic properties of the formation and the hydraulic energy gradients across the flow domain. Groundwater flow rates are generally very slow and are often measured in metres per year. Thus, residence times for a groundwater system may range from months to many hundreds of years. For certain types of media the rate of groundwater flow may be very much higher. In fractured media where the dominant flow is through the fracture network or in Karstic formations containing enlarged conduits, groundwater may travel hundreds or thousands of metres in a few days. The directions of flow are largely dependent on the geometry of the geological formations comprising the aquifer system, their hydraulic characteristics and the topography of the land surface overlying the formations coupled to the available sources of infiltration or recharge water. The groundwater flow distribution within an aquifer may vary substantially over time in response to changes in the natural recharge/discharge distribution and through development of the aquifer system as a source of water supply.

Most frequently, contaminants are present in groundwater in solution and, consequently, are carried with the flowing water. However, as the contaminants pass through the voids of the rock they also come into contact with the rock itself and with other chemical species as well as microbial populations. Thus contaminants are frequently retarded and altered by interactions within the ground-water environment. Such interactions may include adsorption (where the contaminant adheres to the rock surface), precipitation by chemical reaction, decay and reduction by bacterial activity and even volatilization as gases, which are released back to the atmosphere (Bear and Verruijt 1987). Additionally, contaminant concentrations will be changed by dilution and dispersion. Dilution arises from the mixing of the contaminated waters with uncontaminated infil-trating waters. Dispersion occurs because of the tortuous flow paths the water takes through the rock. In general, the hydraulic properties of a rock vary considerably from point to point within a formation. This results in migration of elements of the contaminated water at velocities different from the average rate of movement of the bulk water and causes the development of a mixing front or transition zone. Dispersion occurs both in the direction of groundwater flow and transverse to it. Where a point source of contamination occurs a groundwater contamination plume develops (Fig. 1).

The density of groundwater can be significantly altered by con-taminant concentrations. Even relatively small increases in the density of groundwater may result in alteration of the flow direction of the groundwater and migration of the contaminant plume to the base of the formation. Furthermore, where relatively insoluble contaminants such as oils enter the formation the migration pathways may be substantially different to the natural groundwater flow paths; oil products will exhibit a strong tendency to remain at or near the top of the saturated zone (Fig. 2).

All of the processes of migration and alteration present in groundwater are also present in the unsaturated zone. However, the flow of water through the unsaturated zone is considerably more complex due to the presence of the air and water vapour phases. Nevertheless, it is important to note that attenuation mechanisms in the unsaturated zone can provide a powerful barrier to the passage of contaminants to the saturated zone.

ANALYSIS OF POLLUTION POTENTIAL

Pollution potential can be considered from two standpoints. First, whether a groundwater abstraction zone is liable to contamination and, secondly, whether a potential source of pollution is contaminat-

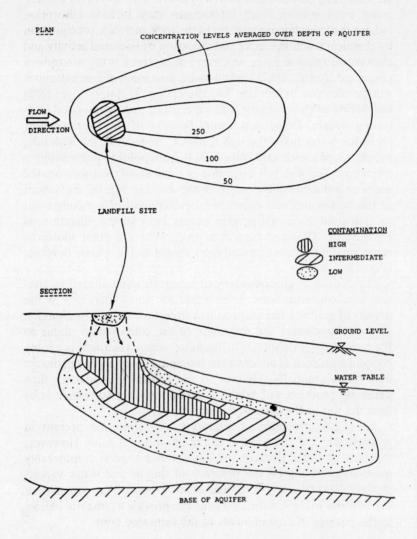

PLAN

CONCENTRATION LEVELS AVERAGED OVER DEPTH OF AQUIFER

FLOW
DIRECTION

250

100

50

LANDFILL SITE

CONTAMINATION

HIGH

INTERMEDIATE

LOW

SECTION

GROUND LEVEL

WATER TABLE

BASE OF AQUIFER

FIGURE 1. Typical leachate plume beneath a landfill site

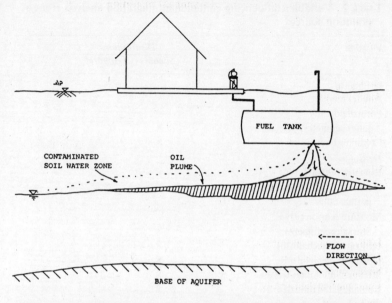

FIGURE 2. Section showing oil pollution plume overlying groundwater

ing the aquifer. In either case the present and/or future conditions applicable to the migration of contaminants must be determined. In the previous section the primary contaminant transport processes within an aquifer system have been identified. In order to be able to use this knowledge for analysis of pollution potential at a particular site, data describing the contaminants, their migration characteristics and the characteristics of the groundwater regime must be collected. Table 2 summarizes the primary variables of interest. The data to be collected can be categorized in terms of:

  (i) Hydrogeological conditions;
 (ii) Contaminant characteristics; and
(iii) Engineering factors.

Engineering factors include the design of waste repositories, water abstraction points and the future programme for the development and use of the groundwater resource under investigation.

Simulation Models

Whilst inspection of the available data can provide a strong insight

TABLE 2. Variables influencing contaminant migration analysis from a pollution source

Variable	Distribution	
	Spatial	Temporal
Geological		
Aquifer media	×	
Form of porosity	×	
Aquifer geometry	×	
Confining beds	×	
Hydraulic		
Effective porosity	×	
Hydraulic conductivity/intrinsic permeability	×	×
Moisture/tension curves		
Storage coefficient	×	
Infiltration mechanisms	×	×
Discharge mechanisms	×	×
Stream/aquifer interactions	×	
Contaminated fluid density	×	
Hydrological		
Precipitation	×	×
Evapo-transpiration	×	×
Surface flow distribution	×	×
Contaminants		
Types	×	×
Adsorption		
Degradation		
Growth		
Dispersion	×	
Alteration		
Anthropogenic		
Demands	×	×
Abstraction designs		
Water supply economics	×	×
Surface water regulation	×	×
Agricultural practice	×	×
Contaminant source management		×

into a potential pollution hazard, the use of models may provide a more appropriate and rigorous method for integrating all the available data together and for evaluation of the response of the aquifer system to a contamination event. Considerable use is now being made of models which allow the prediction of groundwater flow paths and, to a lesser extent, the prediction of contaminant migration pathways (Anderson 1979). These models are generally derived from the expression of the flow and transport processes in terms of mathematical equations, which may then be solved by incorporating appropriate parameter values and boundary conditions derived from the collected field data.

Due to the complexity of most groundwater systems, the solution to the flow and transport problem is generally derived by numerical treatment of the equations. A variety of numerical schemes have been applied to groundwater flow and transport problems using computers and a number of generalized computer codes are now available which can be used for the simulation of most aquifer systems (e.g. USGS-MOC, SUTRA, MOD-3D).

Simulation models may take account of as many of the flow and transport processes in the aquifer as are considered necessary for an accurate representation of the aquifer system. However, although simulation models can be very complex in their formulation, it must be remembered that they remain highly simplified representations of the true aquifer system. In general, they only yield information with a level of confidence dependent on the availability and quality of the field data and the skilful interpretation of the results by an experienced modeller with a strong background in groundwater analysis. Moreover, such models cannot replace data collection if this has not already been carried out or is deemed to be too expensive. Development of a model is an exercise in conceptualizing the true nature of the groundwater regime from the available data and not simply of the generation of numbers by a computer code. A flow chart describing the development of simulation model is shown in Fig. 3.

Frequently it has been possible to use historical data to calibrate and test the accuracy of model for the problem of groundwater flow simulation. Unfortunately this luxury is rarely available for the more complex task of simulating contaminant transport. Consequently, the reliability of results from contaminant transport models are more difficult to evaluate. Additionally, difficulties arise in the modelling of contaminant transport due to more complex problem of collecting data on the parameters such as dispersion coefficients (which define the spread of a contaminant in the

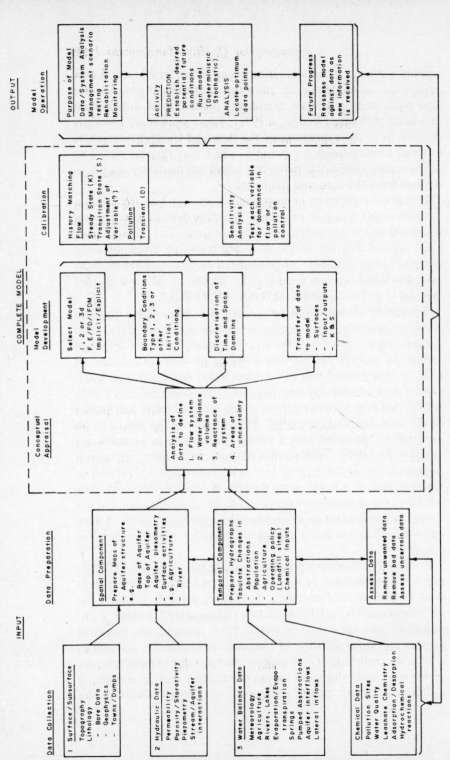

FIGURE 3. Numerical modelling schemes

INPUT

Data Collection

1 **Surface/Subsurface**
Topography
Lithology
- Bore Data
- Geophysics
- Towns/Dumps

2 **Hydraulic Data**
Permeability
Porosity/Storativity
Piezometry
Stream/Aquifer interactions

3 **Water Balance Data**
Meteorology
Agriculture
Rivers, Lakes
Evaporation/Evapo-transpiration
Springs
Pumped Abstractions
Aquifer interflows
Lateral inflows

Chemical Data
Pollution Sites
Water Quality
Leachate Chemistry
Adsorption/Desorption
Hydrochemical reactions

Data Preparation

Spatial Component
Prepare Maps of
- Aquifer structure
e.g.
- Base of Aquifer
- Top of Aquifer
- Aquifer piezometry
- Surface activities
e.g. Agriculture
- River

Temporal Components
Prepare Hydrographs
Tabulate Changes in
- Abstractions
- Population
- Agriculture
- Operating policy (Landfill sites)
- Chemical inputs

Assess Data
Remove unwanted data
Remove bad data
Assess uncertain data

COMPLETE MODEL

Conceptual Appraisal

Analysis of Data to define
1. Flow system
2. Water Balance volumes
3. Reactance of system
4. Areas of uncertainty

Model Development

Select Model
1, 2 or 3d
F. E/FD/IFDM
Implicit/Explicit

Boundary Conditions
Type 1, 2, 3 or other
Initial - Conditions

Discretisation of Time and Space Domains

Transfer of data to model
- Surfaces
- Input/outputs
- K & S

Calibration

History Matching
Flow
Steady State (K)
Transition State (S)
Adjustment of Variable (?)
Pollution
Transient (D)

Sensitivity Analysis
Test each variable for dominance in flow or pollution control.

OUTPUT

Model Operation

Purpose of Model
Data/System Analysis
Management scenario testing
Rehabilitation
Monitoring

Activity
PREDICTION
Establish desired potential future conditions
- Run model (Deterministic
Stochastic).
ANALYSIS
Locate optimum data points

Future Progress
Reassess model against data as new information is received

aquifer) and at a more fundamental level, the quantity of contaminant entering the aquifer.

In spite of many difficulties with models, their real benefit lies in their ability to provide quantified estimates of the potential future states of a groundwater system. The groundwater manager may test, through application of a model, the response of the simulated aquifer system to different options for development of the groundwater regime. Additionally, testing may include the evaluation of the inclusion of contamination sources or the impact of changes to the hydrological cycle. Assuming that the model can be considered to adequately represent the real aquifer system, the manager can effectively distinguish between the different management choices in terms of simulated impacts of each choice of the groundwater regime.

The literature contains a wide range of studies demonstrating the use of simulation models to both groundwater flow and contaminant migration studies (Frind *et al.* 1985; Gray *et al.* 1983; Mackay *et al.* 1988; Mills *et al.* 1986). However, it must be pointed out that by far the majority of studies presented in the literature have been related to research programmes and less to real management application. It is unfortunate that most real applications undertaken by consultants are not presented in the open literature. Moreover, while consultants are frequently employed to develop and use a groundwater model for analysing a particular problem, e.g. wellfield location, groundwater resource yield evaluation, it is rare, in the author's experience, for the client to take over these models and use them for continuous monitoring and appraisal of the groundwater resources of a region. The apparent reticence to use on a regular basis models for management must be overcome if the full advantage of the simulation models applied to resource management are to be realized. Nevertheless, given the limitation of the models, it must be stressed again that models require experienced personnel who have knowledge of both simulation methods and groundwater systems in order to maximize the benefit of a modelling exercise. Extensive training is therefore of primary importance in this field.

Monitoring

Monitoring is an essential element of the maintenance of a groundwater resource. As already indicated, the migration of contaminants in the groundwater regime may extend over many scores of years. Therefore, unobserved long-term contamination may lead to extensive damage to the aquifer which could persist for a considerable period of time. However, regional groundwater quality monitoring programmes for an aquifer system cannot be carried out solely using the same type of monitoring networks adopted for establishing the storage changes in an aquifer. Such a monitoring network would be ineffective in all groundwater quality changes particularly from point source emissions. Monitoring for changes in groundwater quality characteristics is usually much more specific in terms of its location with respect of water supply points or to potential pollution sources.

Groundwater quality monitoring can be separated in four classes (Barcelona *et al.* 1983): *ambient, source, case preparation* and *research*. Of significance to the groundwater quality manager are the first three. Although research extends the knowledge base about the effectiveness of selected monitoring methodologies.

Ambient monitoring is used to establish the variations in groundwater quality over the domain and over time. Such monitoring can be carried out using existing water supply wells, as opposed to specially constructed monitoring wells. The data for ambient monitoring should also include sample data routinely obtained to detect significant groundwater quality changes at primary abstraction sources.

Source monitoring arises in response to known potential pollution sources. Until a pollution event is observed the level of monitoring is usually restricted to a small number of carefully sited indicator wells. The aim in using these monitoring points is to screen the groundwater passing through the potential contamination zone to see whether it is contaminated. Once a contamination event is noticed, the monitoring level will normally increase to delineate the potential impact of the leakage on the groundwater regime.

Case preparation monitoring is undertaken in response to an observed contamination event to produce evidence for legal action against a polluting agency. The obvious objective of this type of monitoring is the provision of proof of the source of contamination.

Research monitoring is used to address the problem of identifying the processes of contaminant migration and degradation as well as for testing new monitoring equipment and methodologies and is therefore both highly intensive and highly selective. Although a necessary component of monitoring, it is seldom of immediate interest to a monitoring organization.

In each of the above cases it can be seen that the aim is to provide a monitoring programme tailored to the specific objectives set for the acquisition of quality data. Before implementing a monitoring programme the purpose of the monitoring must be clearly recognized. Failure to define the objectives of the programme will inevitably lead to inadequate data being provided and additional costs being incurred to rectify the problem.

Developing a monitoring programme should take into account a whole range of technical factors:

- observation network design;
- well designs and materials;
- well construction methods;
- sampling procedures;
- choice of groundwater quality measures; and
- sampling frequency.

Network design, well construction design and sampling frequency depend upon the characteristics of the groundwater regime and the contaminant species. Well construction methods and sampling procedures depend upon the accuracy needed for analysis of the groundwater. The choice of groundwater quality measures will depend on the objectives of the sampling programme and the anticipated contaminant species.

A strong body of literature on monitoring network design and sampling has been put together through the United States Environmental Protection Agency, which provides a valuable source of information for establishing the details of monitoring programmes (Todd *et al.* 1976; Barcelona *et al.* 1983; Barcelona *et al.* 1985). However, it is worth mentioning a few points of general note which affect the development of a monitoring programme. Chemical sampling and analysis must be carried out rigorously and with considerable care (Kirchmer 1983). There is much discussion in the literature about the protocols that should be adopted for a sampling programme. However, all authors agree that failure to appreciate the nature of the chemical conditions of the groundwater to be analysed may lead to erroneous results. Moreover, the methods of extracting the samples from the groundwater environment may produce significantly different sample analyses. Sampling procedures adopted for ambient monitoring, for example, may not give comparable results to sampling procedures necessary for source monitoring.

The use of local materials and construction methods for monitoring well drilling and constructon, whilst apparently cost effective, may significantly reduce the value of the monitoring exercise. In general, the materials used in construction will contaminate or interfere with the groundwater. It would be nice if the choice of materials and drilling techniques could be made solely on the basis of technical requirements and not on economic expedience. However, there are many occasions when accuracy can be compromised in favour of economic factors. Such an occasion might arise when the pollutants are known and the sampling is for qualitative comparison of the degree of pollution. Whilst the choice of drilling technologies and materials may be restricted by local availability, the adoption of good drilling practices must be encouraged during the performance of any drilling contract for monitoring well construction.

Chemical analysis can be both time consuming and costly. Therefore, appropriate decisions should be made about the choice of quality measures required for a particular purpose. For ambient monitoring and for source monitoring simple measures of the basic cations and anions, total organic carbon, and total dissolved solids may suffice as general indicators of pollution. Once a pollution event has been observed, more detailed analyses can follow.

Recently, there has been a growing interest in the use of sampling equipment designed to extract samples from the unsaturated zone (Bumb *et al.* 1988; Kirschner and Bloomsburg 1988). The reasoning behind this development is twofold. First, it reduces the uncertainty of observation inherent in the groundwater monitoring networks arising because of heterogeneities in the aquifer media. Secondly, it establishes at the earliest opportunity, the leakage of a contaminant from the source. Unsaturated zone monitoring devices are now being installed beneath new waste repository installations during their construction. One area of concern for such monitoring is the expected life of the equipment. A repository sited over an aquifer may have a life of many decades whilst the monitoring equipment may be rendered ineffective before the end of the first decade. Research and development is progressing in this area.

Groundwater Protection

The ultimate aim of the groundwater manager is the protection of the groundwater resource and, more importantly, the protection of the abstraction zones from contamination. Development of planning to achieve this objective will depend on the regulatory framework that may be implemented, the type of aquifer system, the potential polluting agencies and sources and the rights of existing and future users over the groundwater resource.

Protection Policy

In order to undertake regulation of a groundwater resource, planning policies must be implemented. For the purposes of establishing a framework for legal policing of the resource, the policy document should be set with respect to (Gladwell 1987):

– the origin of groundwater rights;
– groundwater use and users;
– recognition of groundwater users;
– reallocation;
– modification of or abolition of existing rights; and
– setting priorities for water allocation.

Most important is the existence of an operational regulatory institution. If no recognized body exists to handle that policing of a groundwater resource then one must be established at the earliest opportunity. Jackson (1980) notes that in addition to defining a framework for the conditions of use of groundwater, a complementary system should be established with the expressed intention of controlling those activities that have a potential for pollution of groundwater.

It can be seen that the groundwater manager must have access to a considerable body of data in order to be effective in the regulation of a groundwater resource. In addition, the regulatory body must have defined acceptale standards against which to measure whether or not an individual or group is in breach of or abiding by the policy. These standards will include monitoring requirements, restrictions on abstraction, standards for engineered contaminant structures, such as septic tanks, landfills, underground and surface storage tanks. Furthermore, maximum effluent levels can be defined depending on the value of the underlying aquifer and the hydrogeological characteristics of the region. Liability for aquifer damage will need to be established and the basis for defining aquifer damage agreed. Thus the regulatory institution must develop data collection strategies, define policy documents, define methodologies for monitoring, be able to act on breaches of standards effectively, and act as arbitrator in conflicts of interest. This range of capabilities is onerous and care and time must be taken to develop each aspect or regulation to suit a country's or region's particular conditions. It will in general not be possible to make a transition to a comprehensive management system overnight. Again, adequate training of personnel at all levels in the management and evaluation of groundwater systems is of paramount importance.

Since groundwater systems vary considerably in their relative worth from region to region, policies should vary in accordance with the perceived needs of each individual region. A policy will therefore be different from country to country depending on the regulatory infrastructure, the use to which the groundwater is put and the scale of groundwater development.

Policies usually state that intention for maintenance of a resource only and do not cover the management of the resource to achieve the policy aims. Thus, policies can be divided into the following three categories (Henderson 1987):

  (i) Non-degradation policies;
 (ii) Limited degradation policies; and
(iii) Differential protection policies.

Non-degradation policies attempt to protect all groundwater and maintain the existing or natural quality of the resource.

Limited degradation policies accept that to maintain a resource at present standards will be almost unachievable but aim to maintain the quality of the groundwater at as high a level as possible. The problem with this type of policy is 'how clean is clean' and at what point in the groundwater domain should limits for groundwater quality be set in practice.

Differential protection policies aim to establish for each individual resource zone a value on the resource in terms of acceptable levels of contamination and in terms of the present and potential future usage of the resource. Implementing this type of policy requires appraisal of the value of an aquifer or sub-aquifer system in terms of population served, the availability of other sources, the geological character of the region, and potential liability for contaminant infiltration.

Whilst it may be argued that maintaining the quality of a resource at natural levels eliminates mistakes and errors of judgement about the acceptable limits for individual quality parameters, the possibility of implementing a management strategy to effect such a policy is minimal. A policy which cannot be adhered to is not worth declaring. In reality, the most appropriate type of policy is the differential policy. The level at which a differential policy is administered can, however, be varied, depending on the capacity of the regulatory authority to establish and control the development of pollution through aquifer zoning. At the simplest level it may be appropriate to implement bans on further developments overlying aquifers where strong pollution potential exists whilst accepting, in the short term, degradation due to present pollution potential. Mapping of a country's groundwater resources can be readily implemented in terms of the risk of contamination to groundwater and the distribution of contamination sources. This would then define a straightforward database from which to establish protection zones.

Protection Methods

Protection methods can be reduced to four basic types: *contaminant concentration limits*; *engineering design standards*; *abstraction licensing*; and *land use control*.

Groundwater quality guidelines or standards are generally established with regard to specific chemicals and identifiable bacterial populations indicating pollution. Standards can be set to enforce a minimum acceptable level for each pollutant in the groundwater. Enforcement of the standards is established through permits which acknowledge the rights of individuals or groups to discharge wastes, at or below the limits set, into the groundwater environment. Monitoring of the discharges is necessary and exceedance of the limits must trigger actions aimed at reinstating the quality of the discharge. The location of measurement of the quality of the water is as important as the actual standard. Therefore, discharge zones or boundaries are established around a discharge point beyond which the concentration level must be maintained at or below the standard. In order to place the emphasis of maintenance on the polluter, standards are also required to be met with respect to the sampling procedures and record keeping used to monitor the discharge and ambient groundwater quality. Since concentration of a contaminant in groundwater is a function of the dispersal of the pollutant in the groundwater regime, a requirement for discharge planning may also be necessary to ensure that plugs of undiluted contamination do not enter the aquifer system.

Wherever pollutants are stored in proximity to a groundwater resource, adequate protection of the resource must be achieved by ensuring correct design of the facilities. Design standards relate to construction materials, location of the facility, sizing, operating conditions and monitoring equipment. Equally important is the operational lifespan of the facility. If possible, maintenance checks should be enforced towards the end of the lifespan of storage tanks. Replacement of facilities after a set time period may also be considered for enforcement where highly susceptible groundwater sources are at risk. Siting of a landfill site or other potentially hazardous site may be made with due regard to the hydrogeological conditions prevailing in a region to provide maximum protection against groundwater contamination in the event of failure of the engineered barriers. Siting of monitoring wells and equipment capable of registering pollution activity should be considered if the underlying groundwater resource is at risk.

Abstraction licensing is aimed specifically at controlling excessive local abstraction from an aquifer system. Over-abstraction from the aquifer has the potential to lower the quality of the groundwater supply by drawing in lower quality waters from surrounding zones (sea water intrusion) and from lower formations with older more mineralized waters. Moreover, the zone of infiltration around large abstraction zones may create the potential for long range migration of pollutants to the abstraction points, thereby increasing the difficulty of managing the resource. Good well construction and design practices appropriate to the geology of the region should also be implemented to minimize the risk of local pollution from septic tank emissions and pollution infiltration from the surface. The control of groundwater development can be most satisfactorily achieved through the licensing of local drillers and through the granting of licences by local water authorities to users on a first-come first-served basis. Established groundwater rights might effect how authority to control abstractions is implemented.

Land use control is perhaps the most direct method for controlling groundwater pollution. All actions liable to cause groundwater pollution are identified and rights to use the land overlying groundwater sources are restricted. For this method to be appropriate, standards must be defined for land based on the underlying hydrogeological conditions and the potential for migration of contaminants from the land surface to the groundwater and within the groundwater to abstraction points. Recent articles on groundwater quality management have stressed the importance of delineating recharge zones to the aquifer since these zones are at greatest risk.

Another approach to delineating the relative importance of land areas for aquifer contamination potential is the establishment of vulnerability maps. Different land areas are identified and recorded on maps according to the expected time of migration of a contaminant from the ground surface to the water table (i.e. the residence time in the unsaturated zone):

Zone 1—Residence time longer than 20 years
Zone 2—Residence time generally between 1 and 20 years
Zone 3—Residence time generally 1 week to 1 year
Zone 4—Residence time generally less than 1 week
Zone 5—Highly variable distribution of residence times spatially which are not readily differentiable at the scale of mapping.

The generation of such maps coupled with those showing the underlying aquifer system and the levels of exploitation provide the necessary spatial data to define zones requiring groundwater pollution control.

Subsequent refinement of the zones to be controlled (zones 3 and 4 above) can then be achieved by delineating protection zones around abstraction points which define the limit of recharge to the abstraction point. A reduced zone can be implemented which operates on the basis of an acceptable time of travel to an abstraction source which allows due time for degradation, dilution and dispersion to control contaminant concentration levels (Shafer 1987).

Figure 4 shows an example of a borehole protection zone based on a two hundred days time of travel calculation assuming averaged quantities for the aquifer thickness, velocities, and pumping rates.

Protection zones and vulnerability maps are valuable in defining the potential sites for waste disposal facilities. Additionally, they allow protection from non-point sources of pollution such as agricultural fertilizers and pesticides.

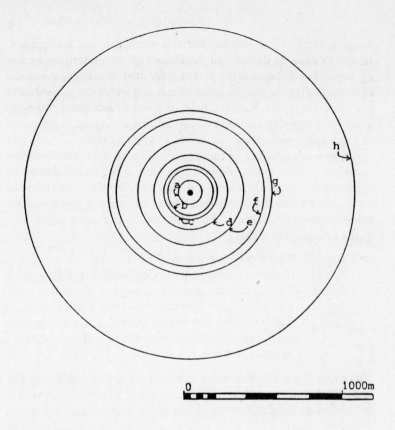

	1	2	
Q -	1000	5000	m³/day (discharge from source)
t -	50	200	days (time of travel to well)
S -	0.1	0.01	- (porosity (eff.))
b -	30		m (aquifer thickness)

Combinations of Q, t, s and b values :

a	=	(1,1,1)	e	=	(1,2,2)
b	=	(1,1,2)	f	=	(2,1,2)
c	=	(1,2,1)	g	=	(2,2,1)
d	=	(2,1,1)	h	=	(2,2,2)

FIGURE 4. Idealized radial flow protection zones (based on times of travel)

REHABILITATION

Once effective pollution of an aquifer has been identified, the first step is to establish a rehabilitation programme. Such a programme can take one of two courses. In the event that a supply point has suffered pollution which is determined to be of limited duration or unavoidable in terms of elimination of the polluting body then treatment of the supply may suffice. If a pollution source is identified which will cause long term damage to the aquifer then rehabilitation of the source and the aquifer may be more appropriate (Sharefkin *et al.* 1984).

Supply Points

When a supply is contaminated, various alternatives may be envisaged for the maintenance of the supply to the end user:

- development of new or existing unpolluted water sources;
- blending of uncontaminated source water with the polluted source water to reduced contamination levels to within the defined safety standards;
- treatment at the supply point; or
- altering the groundwater flow regime to reduce contaminant uptake by the abstraction points.

Development of new or existing unpolluted water source will depend on the availability of such sources. In the case of groundwater supplies the development of adjacent aquifer zone, initially unpolluted, must be carried out with due regard given to the extent of the contaminaton which created the initial problem. If insufficient attention is given to the alteration of the flow paths in the groundwater regime then there may be a risk of migration of the contamination plume into the new supply point.

Blending is only feasible where multiple sources can be mixed to provide an acceptable water quality. However, such an approach requires careful evaluation and monitoring to ensure that the quality of the blended water remains below the specified limits for each quality parameter at all times.

Treatment of contaminated groundwater depends on the type of pollutants present, for example, trubidity, organics, inorganics, bacterial. Standard texts are available on the design of water treatment works for an appraisal of the most common and effective treatments applied to supplies. These include coagulation, filtration, disinfection, gas stripping, etc.

Amendment of the groundwater flow regime can be achieved to reduce the total contamination level entering the source regime by a variety of hydraulic measures. These include:

- reducing pumping rates from each supply point to lower the hydraulic gradients and restrict the ingress of low quality waters from depth;
- lowering the screened section to prevent inflow from the upper aquifer layers if the contamination is restricted to these zones. Such an approach is often effective where oils and organics infiltrating from the surface are the polluting materials;
- establishing recharge wells to dam back the flow of contaminants to the supply point hydraulically;
- establishing contamination abstraction wells to withdraw the contaminated water and thus leave the supply wells withdrawing waters from uncontaminated zones. This approach requires the disposal of the contaminated water produced or its treatment.

The decision about which approach to adopt must depend on the type of contamination, the economics of each method, the technical feasibility of each method and the time available to effect a satisfactory solution. Mixing two or more of the methods outlined above may prove to be more viable than relying on any one technique alone.

Pollution Sites

Rehabilitation of a pollution site, such as a landfill, should be carried out by, first, establishing a recontrol of the site to restrict continued leaching of pollutants into the aquifer and second by examining all cleanup options and implementing the most appropriate method. Recontrol of a site is not a cleanup operation in itself but is merely intended to buy the manager time to evaluate the site properly.

Recontrol actions may include: vacuum extraction of volatile organics; covering the site with an impermeable membrane to prevent infiltration of surface waters; building temporary cutoff walls to keep out surface/subsurface waters and to retain hazardous materials; establishing withdrawal wells to reduce the groundwater flow away from the site; removing the bulk of the wastes and placing them at ground surface where they can be controlled temporarily until

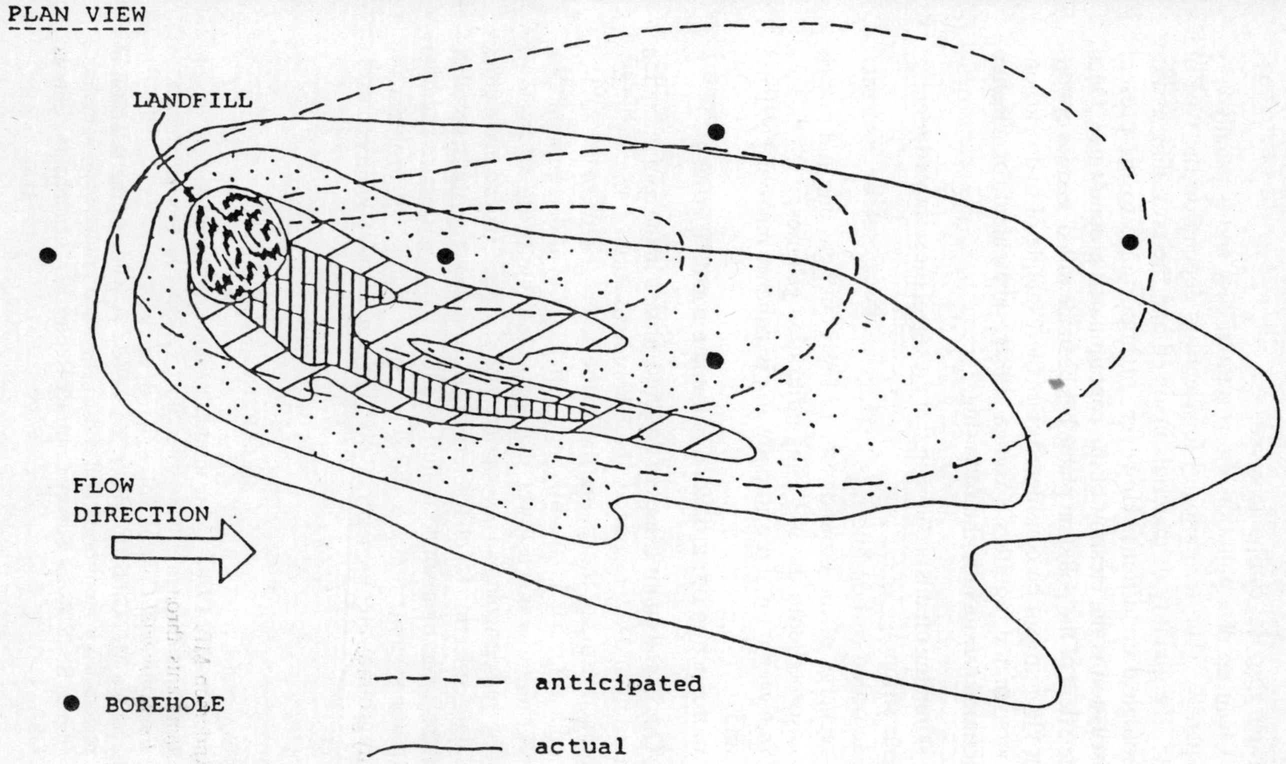

PLAN VIEW

LANDFILL

FLOW
DIRECTION

● BOREHOLE

— — — anticipated

———— actual

FIGURE 5. Effect of heterogeneity on plume geometry

proper disposal can be achieved.

Clean up of a pollution site is seldom easy and is usually very expensive. The technologies for clean up depend on the type of site, the quantity of materials involved and the prevailing hydro-geological conditions. However, of importance to the present discussion is the cleanup of the contaminated groundwater. Once the extent of the pollution plume has been identified, various options for cleanup can be considered. For the example of hydrocarbon-contaminated aquifers (Yaniga 1982), abatement or cleanup methods can include the following:

– skimming off oils accumulated at the top of the saturated zone;
– air stripping;
– activated carbon filtering;
– recharge of the groundwater to flush out light oils and petrols adsorbed onto the soil using infiltration galleries;
– reoxygenation of the groundwater to enhance microbial activity; and
– indroduction of nutrients to stimulate bacterial growth.

One of the major causes of difficulty arising in the contamination of a groundwater zone is the unpredictability of the hydraulic characteristics of the regime due to the heterogeneity of the aquifer media (Fig. 5). Frequently, the efficiency of a decontamination programme is much lower than expected due to the problem of siting and designing an appropriate well network. Indeed, difficulties of data collection for characterizing a groundwater system, whether at the scale of a single pollution site, or over an entire regional aquifer system, remains the single most significant problem for any groundwater manager involved in groundwater quality control.

REFERENCES

Anderson MP. 1979. 'Using models to simulate the movement of con-taminants through groundwater flow systems' in *Critical Review on Environmental Control*, vol. 8, no. 3, pp. 97–156.
Barcelona MJ, Gibb JP, Miller RA. 1983. 'A guide to the selection of materials for monitoring well construction and groundwater sampling', Illinois State Water Survey Contract Report 327. Champaign, Illinois. USA.

Barcelona MJ, Gibb JP, Helfrich JA, Garske EE. 1985. 'Practical guide for groundwater sampling', Illinois State Water Survey Contract Report 374, Champaign, Illinois, USA.

Bear J, Verruijt A. 1987. *Modelling groundwater flow and pollutions*, D. Reidel, Holland.

Bumb AC, McKee CR, Evans RB, Eccles LA. 1988. 'Design of lysimeter leak detector networks for surface impoundments and landfills', *Groundwater Monitoring Review*, vol. 8, no. 2, pp. 102–15.

Frind EO, Matanga GB, Cherry JA. 1985. 'The dual formulation of flow for contaminant transport modelling 2, The Borden Aquifer', *Water Resources Research*, vol. 21, no. 2, pp. 170–82.

Gladwell JS. 1987. 'Groundwater quality management: A multi-dimensional problem' in *The Role of Groundwater Modelling in Decision Making*, Martinus Nijhoff, Holland.

Gray WG, Hoffman JL. 1983. 'A numerical model study of groundwater contamination from Pirce's Landfill, New Jersey, Data base and flow simulation', *Groundwater*, vol. 21, no. 1, pp. 7–14.

Heath RC. 1983. 'Basic groundwater hydrology', United States Geological Survey, Water Supply Paper 2220, North Carolina, USA.

Heath RC, Lehr JH. 1987. 'A new approach to the disposal of solid waste on land', *Groundwater*, vol. 25, no. 3.

Henderson TR. 1987. 'The institutional framework for protecting groundwater in the United States' in *Planning for groundwater protection* (ed. GW Page), Academic Press, New York.

Jackson RE (ed). 1980. 'Aquifer contamination and protection' in *Studies and Reports in Hydrology* no. 30, International Hydrological Programme, 1980, UNESCO, 442pp.

Kirchmer CJ. 1983. 'Quality control in water analyses', *Environmental Science and Technology*, vol. 17, no. 4.

Kirschner FE, Bloomsburg GL. 1988. 'Vadose zone monitoring: An early warning system', *Groundwater Monitoring Review*, vol. 8, no. 2.

LeGrand HE. 1983. *A standardised system for evaluating waste disposal sites*, (Second edition), National Water Well Association, Ohio, USA.

Mackay R, Burgess WG, Cooper TA, Porter JD. 1988. 'Stochastic modelling of solute migration at a shallow waste repository' in Proceedings of the International Symposium on Hydrogeology and Safety of Radioactive and Industrial Hazardous Waste Disposal, Orleans, France.

Mills WB *et al.* 1986. 'Microcomputer analysis of contaminant transport at a superfund site', *Microsoftware for Engineers*, vol. 2, no. 3.

Shafer JM. 1987. 'Reverse pathline calculation of time related capture zones in non-uniform flow', *Groundwater*, vol. 25, no. 3, pp. 283–9.

Sharefkin M, Shechter M, Knneso A. 1984. 'Impacts, costs and techniques for mitigation of contaminated groundwater: A review', *Water Resources Research*, vol. 20, no. 12, pp. 1771–83.

Schepers JS. 1982. 'Use of agricultural BMP's to control groundwater nitrogen', Proceedings of Sixth National Groundwater Quality Symposium, USA.

Todd DK, Tinlin RM, Schmidt KD, Everett LG. 1976. 'Monitoring groundwater quality: Monitoring methodology', U.S. Environmental Protection Agency Report No. EPA/600/4–76–026, Nevada. USA.

World Health Organisation. 1987. 'Guidelines for drinking water quality', World Health Organisation, Geneva.

Yaniga PM. 1982. 'Alternatives in decontamination for hydrocarbon—contaminated aquifers', Proceedings of the Second National Symposium of Aquifer Restoration and Groundwater Monitoring, National Water Well Association, USA.

CHAPTER 9

Environmental Impacts of the High Aswan Dam: A case study

MAHMOUD ABU ZEID

INTRODUCTION

'Egypt', said Herodotus, the father of history, 'is the gift of the Nile.' The importance of the Nile in the survival of Egypt is such that the Egyptian mythology is closely interwoven with the river.

In the long history of Egypt, there have been continuous attempts to build dams across the Nile to retain its water. These were, and still are, prompted by the urgent necessity to increase agricultural production for a rapidly growing population, now estimated at over 54 million.

Of the country's more than one million square kilometres, only 4 per cent is arable. This is concentrated along the thin, fertile strip of the Nile Valley from Aswan to Cairo and in its Delta area on the Mediterranean. Nearly 97 per cent of the poplulation lives along this limited space, about two-thirds of them depending on agriculture for survival. Almost 90 per cent of the country's export earning comes directly from agriculture.

The construction of the High Aswan Dam can be considered to be an irrigation revolution for full utilization of the Nile water. When the Aswan Dam was completed more than twenty years ago, it became a global symbol of environmental and social problems caused by large-scale development projects. A series of papers and articles in scientific and popular media soon made the dam the most popular environmental problem in the world. It was promptly condemned for many reasons, among which were loss of the Mediterranean fishery, increase in schistosomiasis, rising salinity, erosion of Nile bed and banks, reduction in the fertility of the Nile Valley through the absence of the slit deposition, and coastal erosion of the river Delta.

This chapter, written after more than twenty years of operation of the dam, gives the facts based on detailed monitoring and analysis of the positive and negative impacts of the dam by Egyptian and international experts.

REGIME OF THE NILE RIVER

The discharge of the River Nile is subject to wide seasonal variation. About 80 per cent of the total annual discharge is received during the flood season of August to October, the remainder is spread over the rest of the year. About 85 per cent of this annual discharge originates in Ethiopia, and 15 per cent in the lake Plateau in Central Africa. From this plateau rises the White Nile, which is the main source of water to the Nile Valley during the low flow period of the summer. This summer flow was the limiting factor for expanding perennial irrigation in Egypt.

The first attempt to store flood water to augment low flow occurred in 1902 when the old Aswan Dam was constructed. Its capacity initially was 1.0×10^9 m^3. The height of this dam was increased twice to reach a capacity of 5×10^9 m^3 by 1934.

Apart from the seasonal variation in the discharge of the Nile, the total annual discharge is subject to wide fluctuation from one year to another. The highest recorded discharge was in 1878 at 154×10^9 m^3, and the lowest discharge was in 1913 at 42×10^9 m^3. Thus, the short-term subannual storage proved to be inadequate during low flow years for irrigation. The long-term storage became the preferable solution. A suitable location for a long-term reservoir was finally selected at Aswan and the High Dam was constructed to guarantee a constant annual water availability of 84×10^9 m^3 of water, of which 55.5×10^9 m^3 was for Egypt and the balance, 28.5×10^9 m^3, was for Sudan.

Natural Condition of the Nile

The natural annual flow of the River Nile can be divided into two periods: (i) a short 3-month long, high, muddy flow season, and (ii) a longer 9-month low, clear season. At Aswan, the flood period of the River Nile starts usually at the beginning of August and finishes at the end of October. The average annual yield of the river at Aswan is 84 milliard m^3 and the average annual suspended load in the Nile as measured at the tail of the Old Aswan Reservoir is about 134 million tons. The rising and falling period of the high flow is usually accompanied by high suspended quantities of silt from the volcanic highlands of Ethiopia.

The sediment concentration figures vary from 50 to a maximum of about 6600 parts per million (ppm) while the average for the whole flood is about 1500 ppm. This silt is separated by a sedimentation process into sand, silt and clay. For the whole flood, the proportion of the constituents is as follows:

Coarse sand : 0.2 to 2.0 m/m none or trace
Fine sand : 0.02 to 0.2 m/m 30%
Silt : 0.002 to 0.02 m/m 40%
Clay : less than 0.002 m/m 30%

The flood period is divided into rising and falling periods. Generally, the flow in the rising period is accompanied by a high silt concentration. In this period, the analysis of comprehensive field data showed that the mean bed level of the river in the studied reach has increased by about 1.0 m due to siltation and the Manning's coefficient of roughness has decreased to about half its value.

It was concluded that the rising period is characterized by an aggradation trend and creeping of the bed material caused by deposition of the suspended matter, which is higher than the carrying capacity of the river. During the falling period, the mean bed level is further eroded because of the decrease in the concentration of suspended matters. The roughness coefficient 'n' increases again.

It was concluded that the falling period is characterized by a degradation and deformation of the bed resulting from a relative decrease in the suspended matter.

GENERAL DESCRIPTION OF THE ASWAN DAM

The Aswan High Dam is a rockfill dam, constructed across the Nile at a distance of 7 km south of the city of Aswan. The water is diverted into a diversion channel, at the middle of which there are six tunnels. The tunnel inlets are provided with steel gates to control the quantity of water passing through them. Each of the tunnels bifurcates before its end; and each of the twelve branches delivers its water to a hydraulic generating unit of the power station. The diversion channel is situated on the eastern Nile bank. The spillway is located on the western side.

The total length of the dam is 3600 m, of which 520 m is between the two banks of the river, and the rest extend along both sides. The height of the dam is 111 m above the river bed. Its width is

980 m at the bottom and 40 m at the top.

The body of the dam consists of granite and sand, in the middle of which there is a core of Aswan puddle clay to minimize water seepage. At the front, it is connected with a horizontal impervious curtain.

Since the dam bed consists of alluvial materials, it has a vertical cut-off curtain extending below the core, with the same depth as the sedimentary layer, so as to reach the bedrock.

On the downstream side, two rows of vertical relief wells have been provided to drain any water leakage. The reservoir storage capacity is 162×10^9 m^3 distributed as follows:

90×10^9 m^3 : live storage capacity between levels 147 m and 175 m

31×10^9 m^3 : dead storage for sediment deposition

41×10^9 m^3 : storage available for high flood waters between levels of 175 m and 182 m

The hydroelectric power station is situated at the outlets of the tunnels and consists of 12 units with a capacity of 175,000 kW each, i.e. with a total capacity of 2.1 million kW, producing 10×10^9 kWh of power annually. Each generating unit consists of a Francis-type hydraulic turbine.

Work was started on the first stage of the project in 1960 and ended in May 1964. This stage included the excavation of the diversion channel, the construction of the main tunnels, and the construction of the body of the dam up to the level of 132.50 m (47.5 m above river bed). The diversion of the Nile's course to enable the completion of the dam and the power plant took place in May 1964; and the construction was completed in July 1970.

BENEFITS OF THE DAM

In October 1967, the dam body had reached the level of 172 m, and energy was first generated. It is clear that the benefits from the dam started to occur from 1964. From then on, as construction works progressed, the total benefits increased every year. Twenty years have now elapsed since the dam was completed. The following is a brief summary of the benefits based on comprehensive studies:

– improvement of the summer rotations and guaranteed availability of irrigation water at any predetermined period for agricultural production;

- expansion in rice cultivation;
- transfer of about one million acres from seasonal to perennial irrigation;
- agricultural expansion in 1.2 million acres of new land owing to increased water availability;
- protection from high floods as in 1964 and 1975, and from low floods and droughts as in 1972, 1979, 1982, 1985, 1986 and 1987;
- generation of hydroelectric power; and
- improvement of navigation, and further increase in tourism, as a result of the stability of water levels in the Nile course and navigation canals.

Among two other potential benefits are the following:

- Egypt has completed the Toshka flood water storage, about 150 km south of Aswan, to divert excess flood water to the western desert. This project would assist in cultivating some areas of land around the Toshka depression and in the New Valley in the future, when sufficient quantities of water are stored in the depression or in the groundwater aquifers;
- The dam enabled both Egypt and Sudan to begin a comprehensive study for optimal utilization of the Nile waters to increase water resources for the benefit of both countries. The first of such projects is the Jonglei canal, south of Sudan, which is now more than 60 per cent completed.

Many studies have now been carried out on the impacts of the dam on various aspects of the environment.

ENVIRONMENTAL IMPACTS OF THE DAM

Drop in Water Level Downstream the Dam

The Nile Barrages at Esna, Nagga Hammadi and Assiut (Fig. 1) are water level control structures, where gates are used to allow certain discharges through, while maintaining certain required upstrem levels.

Back-water curves upstream of those barrages may extend to tens of kilometres in terms of gradually-varied flow. In all cases, the flow velocity is gradually reduced to a minimum immediately upstream of the barrage. The water passes through two vertical gates above the foundation level in all three barrages.

On the basis of extensive field investigations already carried out,

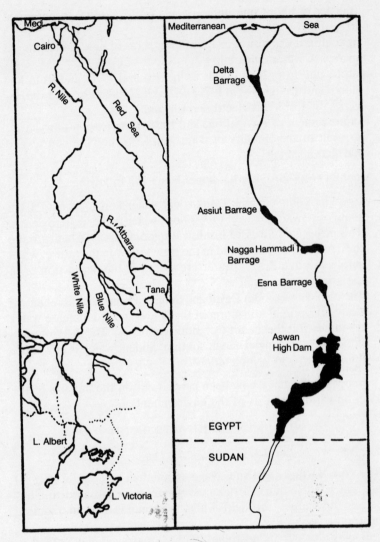

FIGURE 1. The Nile system

it is evident that in each reach of the River Nile, between two barrages, there is a general degradation region followed by an aggradation region with no appreciable amount of bed material moving into the next reach. However, whereas aggradation is taking place due to relatively fine bed material, degradation is accounted for by relatively coarse bad material.

Table 1 shows the actual drop in water levels caused by degradation at four sites along the River Nile from Aswan during the period 1963–86. It is assumed that the year 1963 represents the natural condition of the river.

TABLE 1. Drop in water level (W.L.) caused by degradation

Site	Distance from Aswan Dam km	Drop in W.L. (m) for daily discharge (million m³)			
		80 Mm³	100 Mm³	150 Mm³	200 Mm³
Gaafra	34	0.9	0.76	0.44	0.28
d/s of Esna Barrage	167	1.03	0.95	0.76	0.6
d/s of Nagga Hammadi Barrage	360	0.72	0.92	0.85	0.6
d/s of Assiut Barrage	540	0.74	0.74	0.63	–

d/s downstrem

It is interesting to compare the results of predictions made by various water experts during the past 24 years with the actually observed data of bed degradation. Table 2 summarizes the result of degradation on the drop of water levels along the reach between Aswan and the Delta Barrage. It clearly shows significant variations between what were predicted by the experts, and the actual results.

Rise in River Water Levels

After degradation, the eroded materials move longitudinally downstream and are deposited in the deeper cross sections, which cause water levels to rise in those reaches. This phenomenon was observed at Ekhssas gauging station (68 km upstream of the Delta Barrage). Water level rose by 1.0 m during the period 1963–86.

TABLE 2. Intercomparison of theoretical predictions and observed data in drop of water levels (m)

Reference	d/s H.A.D. (Gaafra)	d/s Esna Barrage	d/s Nagga Hammadi B.	d/s Assiut Barrage
Predictions by:				
A. Fath (1956)				
G. Mostafa (1957)	8.50	8.00	7.00	6.50
V. B. B. (1960)	3.0–4.0	3.0–4.0	3.0–4.0	3.0–4.0
S. Shalash (1965)	2.0–3	2.0–3	2.0–3	2.0–3
D. B. Simons (1965)		3.50		
Hydro Project (1973)	3.00	3.50	3.50	3.00
S. Shalash (1974)	1.37	1.01	1.37	2.90
Hydro Project (1976)	5.30	7.00	11.00	
Hydro Project (1977)	3.00	2.50	4.00	8.00
M. El-Mottassem (1977)	3.2	2.1	2.4	
Observed				
Q = 80 Mm³/day	0.9	1.03	0.72	0.74
Q = 200 Mm³/day	0.28	0.60	0.6	

d/s downstream

Bank Erosion and Meandering

A meandering river such as the Nile has a series of bends. In the bends, deep pools are formed adjacent to the concave banks by the relatively high water velocities, which usually cause bank erosion. Because velocities are lower inside the bends, sediments are deposited in this region, thus forming bars which usually create navigation problems.

Studies have been initiated since 1973 to investigate the erosion levels and failures taking place at certain locations along the channel banks. In each case, soil samples were taken and tested and the problem was analysed to determine the cause of failure and to identify the steps that should be taken to prevent failures at other similar locations.

It was concluded that the total length of the eroded channel banks along the River Nile is about 500 km. This represents more than a quarter of the total length of the river channel bank from Aswan to Cairo (900 km).

The main reasons for the erosion of the Nile banks are the following:

1. meandering of the river;
2. groundwater seepage from the bank;
3. seepage of water from adjacent lands towards the river;
4. wave action due to navigation;
5. reduction of the water levels after the construction of the Aswan Dam; and
6. reduction of the river slope due to degradation.

The Downstream Scour Holes

Scour holes exist downstream of the existing barrages at Esna, Nagga Hammadi and Assiut. They are not located over the whole width of the river. At Esna and Nagga Hammadi barrages, there are two scour holes at the eastern and western sides of the river.

According to the periodic hydrographic surveys of these scour holes from 1973 to 1982, it has been observed that their volumes had increased by about 30 per cent under a constant datum. It was recommended that the scour holes be treated by filling them with rip-rap to be placed over an inverted filter, designed according to standard specifications.

Water Quality after Construction of the Dam

Before construction of the Aswan Dam, there were seasonal changes in certain water quality constituents of the Nile water. During the flood period, the Nile water is muddy, it has minimum dissolved solids but maximum suspended solids; whereas during the rest of the year, dissolved solids increase but suspended matters decrease.

After construction of the dam, the water released from the reservoir is practically silt-free, and the maximum discharge is about a quarter of the earlier flood discharge. In addition to the reduction of discharge, both the population and industrialization of the country have increased significantly during the past two decades. It is thus very necessary to assess the quality of the Nile waters for domestic, agricultural and industrial uses.

Some water quality monitoring has been carried out since 1976. Water quality samples are taken every 10 km from Aswan to the Mediterranean Sea and also within 200 metres upstream and downstream of the major discharge points. In addition, one sample is taken from the discharge point, where all the water quality variables are measured. The flow is also measured at the sampling point. Chemical and biological analyses are carried out. Some 86 point sources were monitored between the Aswan Dam and the Delta Barrage, of which 20 were liquid industrial wastes and 66 were agricultural drains.

In 1986, the total annual discharge from the point sources monitored to the river, between Aswan and Cairo, was estimated at 3.6 milliard m³, of which 3.2 milliard m³/year was from agricultural activities and the balance, 0.40 milliard m³/year, originated from industry.

An earlier study on the variation of water quality of the River Nile, from Aswan to Cairo during the period 1976 to 1985, showed that there was significant deterioration in terms of the Water Quality Index. The following parameters were used to estimate the index: temperature, pH, dissolved oxygen, biochemical oxygen demand, total dissolved solids, suspended matters, phosphates, nitrates, ammonia and faecal coliforms.

The index had a lower value in 1985 as compared to 1976, which indicated that the water quality deteriorated owing to increased industrial and agricultural discharges. It was also found that the index is affected by water levels upstream of the dam. The index was high for a high water level and vice versa, which means that the more water stored in Lake Nasser, higher is the water quality. This indicates that the water pollution control law (No. 48 of 1982) must be more strictly applied and all waste discharges to the river must be treated properly.

Siltation in the Reservoir

The mean annual sediment load of the Nile at Aswan has been estimated at 134 million tons. Siltation in the reservoir generally depends on the volume of the flood water, and also on the distribution of the flood water within the reservoir.

When the reservoir was in the process of being filled, silt deposition covered wider areas depending on the size of the floor. More recently, siltation has been confined to the most southern 200 km of the reservoir. Particularly heavy siltation occurs in the area between 360 and 430 km south of the dam. In this region, the level of siltation has already affected the life storage zone (Fig. 2). Nevertheless, sediment transport phenomena tend to be in a northerly direction.

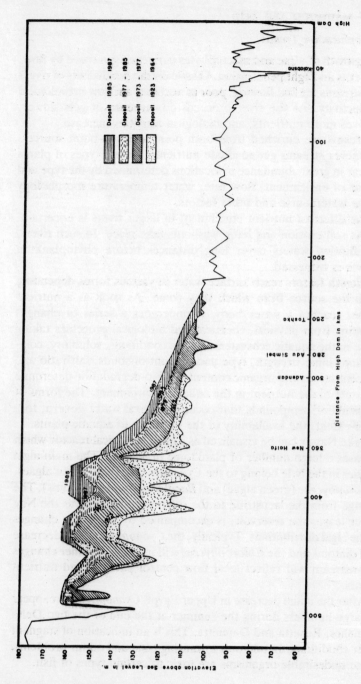

FIGURE 2. Siltation in Aswan High Dam Reservoir

Eutrophication Trends

The growth of algae and macrophytes in rivers is governed by flow, nutrients and light penetration. Classically the headwaters of rivers and streams are fast flowing, poor in nutrients and low in biological productivity. As the river proceeds downstream, it gets slower, receives more nutrients, and biological activities increase.

Streams are enriched from both point and non-point sources. Whenever streams get adequate nutrients, certain types of plants appear in great abundance in locations determined by the type and extent of enrichment, flow rate, water temperature morphology of the watercourse and other factors.

The effect of nutrient enrichment in larger rivers is uncertain, unless soil erosion and bank washouts take place. In such rivers, fast flowing waters cover long distances before phytoplankton growth is witnessed.

Growth factors reach surface water in various forms depending upon the source from which they come. As soon as a nutrient carrier reaches a water body, it undergoes a series of changes resulting from physical, chemical and biological processes taking place in the aquatic ecosystem. Redox conditions, solubility, complexion, ionic strength, type and amount of solids, salt ratio and concentrations and organic content and bio-degradation determine the state of the nutrient in the aquatic environment. The forms of elements or compounds that occur in natural water govern, to a large extent, the availability of the nutrients to aquatic plants.

Lake Nasser can be visualized as a huge biological reactor which releases a large number of planktonic organisms. The main algal species in the Nile belong to the *Cyanophyceae* (blue-green algae), *Chlorophyceae* (green algae) and *Bacillariophyceae* (diatoms). The change from the lacustrine to the riverine conditions, as the Nile water leaves the reservoir, is accompanied by significant changes in the algal distributions. Typically, the *Cyanophyceae* will decrease and diatoms and the *Chlorophyceae* will increase. Further changes downstream will reflect local flow control patterns and nutrient inputs.

After the initial decrease in Upper Egypt, *Cyanophyceae* reappear in large numbers during the summer at the end of the two Delta branches, Rosetta and Damietta. This is an indication of stagnant river conditions. In addition to causing taste and odour problems, these undesirable organisms are toxic to certain types of fish.

Estimation of Evaporation from the Reservoir

The maximum width of Lake Nasser is about 60 km, with an average width of about 10 km. It has a surface area of about 5000 km^2 at the 175 m water level. The average depth of the lake is 25 m and the maximum depth is 90 m. The lake is surrounded by desert and hilly land. It lies in a hot and extremely arid climate. Four meteorological stations were constructed at the dam site and along the lake. The main climatological factors at the meteorological station 3 km west of the High Dam are as follows:

– Mean maximum air temperature varies from 23.5 °C in January to 41.8 °C in June.
– Mean minimum air temperature varies from 8.1 °C in January to 24.8 °C in July.
– Mean relative humidity varies from 13 per cent in May and June to 37 per cent in December.
– Mean wind speed at 2 m height varies from 7.8 km (4 m/sec) in December to 9.3 km (4.7 m/sec) in April. The prevailing wind direction is northerly.

Several studies were carried out to estimate the evaporation from the lake. In a recent study, the average monthly evaporation was estimated by the heat budget and bulk aerodynamic approaches, using monthly estimations of different meteorological elements over the lake.

Table 3 shows the mean daily values of evaporation calculated on this basis. It shows that the mean daily evaporation is maximum in June (10.9 mm/day), and minimum in January (3.8 mm/day). The mean daily value of evaporation for the year as a whole is 7.35 mm/day. Highest evaporation during a period of four consecutive months occurs between June and September when evaporation accounts for 45 per cent of the total yearly value.

The monthly total water losses by evaporation are proportional to the rate of evaporation and the lake surface area, which is maximum during the flooding season. Maximum losses occur from June to September. The total annual losses at the average lake level of 175 m above mean sea level is estimated at 13.7×10^9 m^3, which is about 11 per cent of the water stored.

Coastal Erosion

The formation of the Nile Delta is due to the direct result of silt

TABLE 3. Mean daily values of evaporation calculated by the heat budget method (E_H) and the bulk aerodynamic methods (E_B)

Months	(E_H) (mm/day)	(E_B) (mm/day)
January	3.59	3.93
February	4.95	4.08
March	5.40	4.77
April	5.52	7.32
May	8.95	9.31
June	11.66	10.11
July	10.42	10.01
August	8.39	10.68
September	8.61	10.47
October	6.98	7.83
November	4.87	5.17
December	4.90	4.52
Mean	7.35	7.35

deposition. During the prehistoric period, the Nile had several mouths to the Mediterranean Sea. The northern coast has assumed its present shape since ancient times. This is indicated by the lakes (Edku, Burullus and Manzala) between and on the sides of the two remaining branches of the Nile. A quasidynamic equilibrium was established between silt deposition and wave actions. Observations made since 1898 indicate the erosion of the delta in some locations at Rosetta and Damietta.

Some scientists believe that recent erosion has resulted from the change of the Nile regime caused by construction of the Aswan Dam, which prevents silt releases to the Mediterranean. This is now being seriously investigated. It should, however, be noted that some 15 years before the dam was constructed, the problem of coastal erosion was of great concern to the Egyptian ports and lighthouse authorities. In 1930, coastal engineering works were built on the central hinterlands of the delta to protect the relocated village of Burg El-Burullus. In 1940, the famous jetty of west Damietta was built. These and other protective structures, erected at different places along the coast, worked to some extent until the construction of the Aswan Dam. Since then, the rate of erosion has increased. The silt that used to be carried every year to the Rosetta estuary (about 90 million tons) and to the Damietta (about

30 million tons) has now decreased considerably. This annual sedimentation compensated the erosion due to the effects of sea currents and wave action. Studies of all the elements affecting the new regime of the Nile and the Delta coastline were started in 1970 and are continuing. Enough data are available to recommend important protective measures.

In 1985, a comprehensive coastal protection plan was outlined and approved by an international panel of experts. Some of the protective works were implemented in 1986, namely, above Abu-Quir and Rasheed protective walls.

Groundwater Regime

The Nile Valley and the Delta, from Aswan to the Mediterranean, cover an area of about 3 million ha. Within this region, there is a groundwater aquifer which is hydrogeologically connected to the Nile. The aquifer system consists of graded sand and gravel, overlain by a semi-pervious layer of silty-clay.

Before construction of the Aswan Dam, the river flood was the major factor affecting groundwater levels in the Nile Valley and Delta aquifer. Groundwater levels used to rise after the flood wave, but with a reduced amplitude and a time lag depending upon the distance from the Nile. These levels then used to decrease gradually, reaching a minimum in June, just before the next flood.

Due to the difference of levels between the Nile and groundwater, an exchange of water used to take place. In general, the Nile was the main source of groundwater replenishment. Vertical seepage also occurred due to irrigation and natural drainage used to take place laterally to the Nile during low flow periods.

After the dam was constructed, water releases were according to requirements, and the river water levels are very similar all through the year. As a result, the cyclic behaviour of the groundwater levels was greatly reduced. However, with the increasing cropping intensity and expansion of perennial irrigation, more irrigation water is now being released, which has increased vertical seepage. The lack of effective drainage in some areas of the valley and delta has resulted in a continual increase in both shallow and deep groundwater levels. Figures 3 and 4 show two typical groundwater hydrographs in the Delta and Nile Valley.

Figure 5 shows an iso-salinity contour map of the Delta aquifer for 1978. The iso-line of Revelle ratio which is equal to 250 is

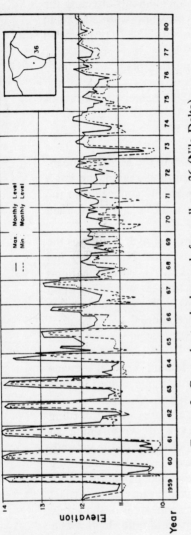

FIGURE 3. Groundwater hydrograph for well no. 36 (Nile Delta)

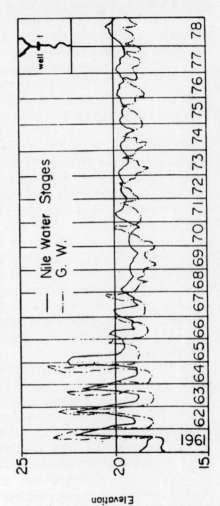

FIGURE 4. Groundwater hydrograph for well no. 1 (Nile Valley)

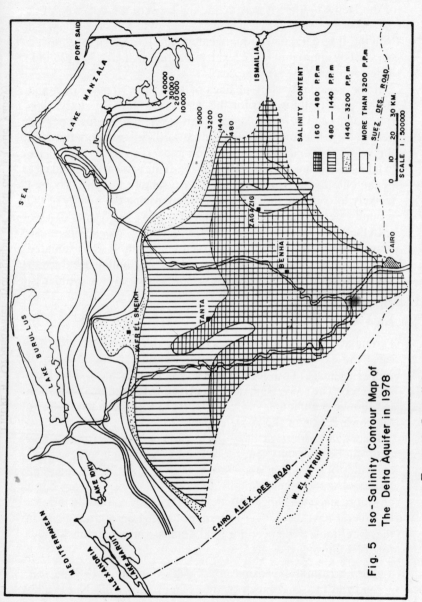

Fig. 5   Iso-Salinity Contour Map of
The Delta Aquifer in 1978

FIGURE 5.   Iso-salinity contour map of the delta aquifer in 1978

considered an index of the Mediterranean water in the coastal part of the Delta aquifer. It may define the locus of the upper point of the sea water wedge intrusion in the aquifer. When this was compared with the 1968 data, it was found that within 10 years this line moved seaward by approximately 2.5 km, indicating an increase in groundwater gradients towards the sea.

If the iso-salinity contour lines of 1000 ppm (fresh water) and 30,000 ppm (salty water) for the years 1958 (flood period), 1966 (after dam) and 1978 are compared, it is evident that for the flood period, these lines moved inland and seaward with the flood. Between the years 1966 and 1978, a transition appears to have taken place in terms of continual seaward movement. If the present water balance of the Nile Delta is unchanged, the iso-salinity line of 1000 ppm is likely to continue moving seaward, improving the quality of the groundwater in the southern Delta.

In the Lake Nasser area, on the basis of hydrogeological investigations carried out before the construction of dam, it was estimated that annual seepage losses would be around 750 million m^3. Current studies indicate that seepage losses are within the earliest estimate.

Drainage Before and After the High Aswan Dam

Perennial irrigation has been used in Egypt for multiple cropping since the early part of this century. Because of the absence of proper drainage, both water table and salinity in many areas have increased. The conversion of basin irrigation to perennial irrigation for about 0.5 million ha in Upper Egypt took place in parallel with the construction of the Aswan Dam.

Inadequate drainage has been a problem since 1938. Salinity and waterlogging problems developed even though some natural leaching took place in many areas close to the Nile channel and branches as the annual flood water receded. This does not occur any more since the Nile levels are now relatively stable and farmers in general have been over-irrigating, causing an alarming rise in the water table. The need for field drainage became more serious for the following reasons:

1. the horizontal expansion in agriculture in sandy or light soil, and the fact that much of these lands lie on the Nile Valley fringes of higher elevations;

2. seepage from new irrigation systems;
3. increased cropping intensities from one to two crops a year; and
4. increase in rice and sugarcane growing areas, requiring higher water applications.

Table 4 shows the relation between irrigation and drainage water for the period 1960 to 1986.

The total area requiring sub-surface drainage is estimated at 2.1 million ha. Before 1960, only 100,000 ha were provided with tile drains. It is expected that all of 2.1 million ha will have sub-surface drains by the end of 1991. The cost of field drainage system is being repaid by the farmers on 20 year interest-free annual instalments, starting by the third year after completion of implementation.

Social Impacts

The Aswan region, which has remained dormant for centuries but was an important centre during the early kingdoms, has become a focus of economic activity since the construction of the dam.

TABLE 4. Irrigation and drainage water before and after the High Aswan Dam

Year	Irrigation water (million m³)	Drainage water (million m³)	%
1960	23.459	10.658	45
1961	21.925	10.497	48
1962	26.053	12.623	49
1963	26.735	13.597	51
1964	26.990	14.025	52
1965	30.054	14.386	49.5
1966	32.003	14.847	46.5
1967	29.173	16.020	54
1968	29.731	15.528	52
1969	30.246	15.941	53
1970	31.178	16.238	52.5
1971	32.175	16.261	50.5
1972	32.591	15.989	50
1980–3	34.980	16.418	46
1984	36.097	16.840	46
1985	34.505	16.501	47
1986	34.929	16.415	46

With the establishment of large-scale industries, such as fertilizer, iron and steel, sugar, ply wood, milk industries, pulp and paper, as well as phosphate, kaoline and other mines, new population centres have emerged around the dam area. The population of the Aswan area has increased from 280,000 in 1960 to around one million, mainly becasue of the increase in job opportunities.

Furthermore, the deserted Nubia, flooded by the reservoir, has begun to flourish again. In Abu-Simbel, the rescue of the two famous temples of Ramsis II has transformed the area into a major tourist centre.

Fish Production

Extensive studies have been conducted on fish production in Lake Nasser. Fish distribution and abundance varies among the different sections of the reservoir. Furthermore, different khores frequently exhibit different fish distribution. The migration of certain types of fish has been dependent on the arrival of the turbid floodwater and their preference to riverine or semi-riverine conditions. Fish behaviour partially accounts for the difference in their contribution in catches, which is quite significant for apecies of economic importance.

The Tilapia species are the most predominant in fish landings, followed by *Lates niloticus*, Catfishes and Cyrinds represented by *Labeo* and *Barbus* species. Recently, *Tilapia galilaea*, which was sporadically recorded in the early years of the impoundment, has become the most abundant species in fish landings.

Sardines which breed at the estuaries of the Nile have almost disappeared. The mineral rich silts which nourished the sardines are now deposited behind the High Dam. The annual catch of sardines has dwindled from 80,000 to 60,000 tons annually.

The HAD Lake Authority, charged with developing the potential of the lake and its environs, is now encouraging former residents to return to the area to help develop the fishing and farming industry. Fish production through hatcheries and intensive fish farming is seen as having a major potential. Already there are about 7000 fishermen on the lake. The main problem is industry and marketing.

Effect of the High Aswan Dam
on Soil Fertility

Some researchers have claimed that the trapping of the sediments in Lake Nasser will affect the fertility of old lands, and thus chemical fertilizers have to be used to maintain soil fertility. Studies indicate that 88 per cent of the Nile's annual sediment load used to go to the Mediterranean. It was estimated that only 6000 tons of potash (K_2O), 7000 tons of phosphorous pentoxide (P_2O_5) and 17,000 tons of nitrogen (N) were added to the land annually. In relation to the fertilizer need for large yields, these amounts are insignificant.

More significant to the fertility requirement of the land is the increase in yield through reliable supply of irrigation water, the growing of two or more crops per year, and the resulting high levels of production with consequent increases in withdrawals from the soil. These withdrawals require replacement and supply of adequate amounts of the major nutrients like nitrogen, phosphorous and potassium. This is probably a major contributing factor to the need for a number of trace elements in recent years. The continuing nutrient requirements for increased production has an effect many times higher than due to the sediment loss.

Studies have indicated too that there is no direct cause and effect relationship between the 'High Dam and the lack of certain trace elements. There is no direct evidence that trace elements were present in adequate quantities in the soil prior to the High Dam and that the deficiency did not exist prior to its construction. Furthermore, trace elements can be easily and economically added to the soil.

Effect on Historical Monuments

Lake Nasser did inundate some monuments but the major ones, Abu-Simbel and Kalabsha Temples, were saved by international efforts. These were reconstructed on higher elevations. UNESCO organized a resuce team which accomplished one of the greatest engineering feats of recent times. Engineers, architects and archaeologists from many nations participated in saving monuments which were built by the ancient Egyptians along the Nile. Several temples were in fact relocated, but the biggest challenge was the two temples at Abu-Simbel cut into the solid rock. The statues, columns and frescoes were systematically cut into blocks, numbered, lettered and transported safely to a plateau above the level of the lake. The 1050 blocks were reassembled without a single block being broken or cracked. The joints have been made so carefully that they are almost undetectable.

There was some concern that the increase in groundwater levels, waterlogging, and salinity would damage the monuments downstream, or that the increase in groundwater would decrease the bearing capacity of the soil and wells of temples which might collapse. the higher water table would allow salts to move up into the monuments by capillary action. Both these actions have occurred in the past and could occur in the future.

However, it should be noted that during floods, temples and monuments were flooded. More research is now being undertaken on the action of the water on the monuments and efforts needed to preserve them. However, it appears unlikely that the High Dam had a significant adverse effect on the monuments downstream.

Health Impacts

Since the construction of the dam, there have been a number of plans drawn by several agencies for economic development of the region along the shores of the reservoir. These include the establishment of human settlements (e.g. irrigation, land reclamation), industrial development, and tourism and recreational projects.

An important risk is that such development schemes will affect water quality and the aquatic ecosystem of the reservoir. Furthermore, the deterioration of the quality of the reservoir will affect downstream river uses.

A serious consequence could be the propagation of schistosomiases and the northward migration of malaria mosquito vectors from Sudan. The *Anopheles gambiae* has invaded Egypt on several occasions, the last of which was in 1942. The vectors of both *Schistosome mansoni* and *Schistosome haematobium* have already been found in Lake Nasser. The assumption is that the fishermen brought the disease with them to the region.

Policy-makers should consider the alternative of keeping the reservoir area largely free from human activities, i.e. restricted and protected area, in an attempt to avoid possible pollution problems and the spread of water-brone diseases. Alternative plans for the development of the shores of the reservoir should incorporate more realistic assessments of the environmental impacts and the cost of pollution control.

CONCLUSION

In retrospect, an objective evaluation of the impact of the Aswan Dam based on more than twenty years of operational data clearly indicates that it has overall been overwhelmingly positive, even though it has contributed to some environmental problems. These problems, however, are significantly less than most people originally feared. As the Executive Director of the United Nations Environment Programme, Dr Mostafa Kamal Tolba, said at the recent World Congress of the International Water Resources Association in Ottawa, 'the real question is not whether the Egyptians should have built the Aswan Dam or not—for Egypt realistically had no choice—but what steps should have been taken to reduce the adverse environmental impacts to a minimum'. It is clear that after more than twenty years of operation, the Aswan Dam now deserves much more credit than it has received so far for beneficial contributions to Egypt's overall development.

REFERENCES

Abdel-Latif A. 1981. 'Management of fish resources of Aswan High Dam reservoir', Academy of Scientific Research and Technology, Cairo, Egypt.
Abu El-Atta A. 1978. 'Egypt and the Nile after the High Dam', Ministry of Irrigation, Cairo, Egypt.
Abu Zeid M. 1969. *Drainage in Egypt*, Ministry of Irrigation, Cairo, Egypt.
———. 1971. *Groundwater in Upper Egypt*, Water Research Centre, Egypt.
———. 1971a. 'Prediction of groundwater levels in the Nile Delta area after Aswan High Dam', Ministry of Irrigation, Cairo, Egypt.
———. 1978. 'Short- and long-term impacts of the River Nile projects', Blue Plan Seminar on Soft Water Problems in the Mediterranean Zone, France.
———. 1980. 'The River Nile, its main water transfer projects in Egypt and its impacts on Egyptian agriculture', Task Force on the Transfer of Water from Yangtse River to the North China Plains, Beijing, China.
El-Gohary F. 1981. 'Water quality changes in River Nile and impacts of waste discharge', Environmental Division, National Research Centre, Cairo, Egypt.
Fayed S. 1981. 'Eutrophication trends in the River Nile', Egyptian Academy of Scientific Research and Technology, Cairo, Egypt.
Hammad HY. 1981. *Losses from the High Dam reservoir*, Faculty of Engineering, University of Alexandria, Egypt.
Henderson J. 1973. 'Actual and potential yield of fish in the High Dam lake', FAO, Report (FI, SF/DP, EGY/66/558/11).
Kotb M, Abdel-Mottaleb F. 1981. 'Effect of High Aswan Dam on the regime of the river downstream Esna Barrage', High Aswan Dam Side Effects Research Institute, Cairo, Egypt.

Lasheen M. 1981. 'Selected trace elements in Aswan High Dam reservoir and River Nile ecosystems', Water Pollution Control, National Research Centre, Cairo, Egypt.

Latif A. 1978.'Effect of impoundment on Nile Biota in Lake Nasser', Proceedings of International Symposium on Environmental Effects of Hydraulic Engineering Works, Tennessee, USA.

Mahmoud M, Guariso G, Younis M. 1979. 'Lake ecosystem modelling, application on High Dam lake', Egyptian Academy of Scientific Research and Technology, Egypt.

Mancy K, Hafez M. 1977. 'Water quality and ecosystem considerations in intergrated Nile resources management', Conference on Water Resources Planning in Egypt, Ministry of Irrigation, Cairo, Egypt.

Younis M, Mancy K. 1981. 'Model development and optimization of the River Nile water quality', Egyptian Academy of Scientific Research and Technology, Cairo, Egypt.

———. 1981a. 'Water quality data bank', Egyptian Academy of Scientific Research and Technology, Cairo, Egypt.

Index

274 / *Index*

research 37, 56, 102–18
resettlement 24–5
resilience 33
risk assessment 67–8, 76, 103
River
  Amazon 3, 158
  Brahmaputra 158, 167
  Chiang Jiang 158
  Colorado 160
  Columbia 34
  Congo 3, 4, 18
  Euphrates 6, 15, 158
  Ganges 158, 167
  Hai (He) 157, 158
  Huang (He) 6, 158, 159, 163, 170
  Indus 6, 15, 158
  Meghna 167
  Mississippi 158
  Nile 13, 17, 18, 43, 158, 159, 160, 247–70
  Niger 17, 159, 165
  Tigris 6, 15, 158
  Wuding 170
  Zaire 157, 158, 159
  Zambezi 24
Roser, S. P. 7, 29

Sagardoy, J. A. 122, 140
Sahel 161
salinity 9, 12, 13, 14, 15, 47, 54, 132, 134, 186, 247, 261, 264, 267
salt water intrusion 9, 13, 223, 238, 264
sampling error 124, 128
schistosomiasis 19–20, 47, 51, 52, 53, 247, 268
sedimentation 10, 12, 17, 22, 47, 144, 155, 157–67, 181, 186, 189, 190, 202, 249, 256–7, 267
seepage 5, 15, 128, 255, 261, 264
seismicity 9, 15–16
sensitivity analysis 67
Sharma, V. P. 51, 58
Shuval, H. I. 211, 218
Sierra Leone 43
simulation models 76–8, 83, 114, 202–3, 227–31
small-scale projects 56, 75, 154, 169–73
social impacts 42–5, 47, 265–6

soil fertility 14, 22, 266–7
soil loss 13
Sokoto 161–2
Sri Lanka 6, 23, 170
stabilization pond 208, 211, 215
Steinberg, D. J. 122, 140
Sterling, C. 47, 49
stochastic models 202
Stockholm 46
Streeter–Phelps equation 200, 202
streptococci 179–80
sub-Sahara 66
Sudan 20, 156, 268
Sundborg, A. 12, 18, 29
Surinam 12, 18

Taiwan 209
Tam, D. M. 1–29, 176–219
Tanzania 43, 97
technology 26
  appropriate, 22, 108, 113, 116, 215
  labour-intensive, 10, 134
  transfer, 153–4, 161
temperature 1, 2, 8, 9, 11, 16, 178–9, 187, 190, 191, 198, 203, 204
Tennessee Valley 40
Thailand 14, 17, 19, 23, 24, 97, 104, 152, 157, 189, 208, 209, 210
Thanh, N. C. 1–29, 176–219
thermal pollution 108, 178–9
thermal stratification 11, 17, 190, 204
Todd, D. K. 233, 246
Tolba, M. K. 32, 58, 269
toxicity 17, 34, 180–1, 187, 188, 195
trace elements 267
training 33, 56, 92–3, 102–18, 149, 151, 154, 214
trihalomethane 180
trypanosomiasis 21, 25
turbidity 2, 12, 187, 188, 190, 198, 203, 204, 241, 266
turbulence 204
Tuntawiroon, N. 14, 29

uncertainties 72, 84
United Kingdom 43, 44, 45, 107, 110, 111, 112
USSR 158
United Nations 31, 46, 51, 116
  Conference on Human Environment 46